The Gen AI Manufacturing Revolution

The Gen AI Manufacturing Revolution

Smarter Factories, Enhanced Products, and Reduced Costs

Matthew Alberts, PhD

WILEY

For my wife, Kelli, and my three daughters, Brooklyn, Milena, and Blakely

Nothing is out of reach

About the Author

 **Matthew Alberts, PhD**, is a technology strategist, engineer, and thought leader in the field of innovation and emerging technologies. With over 15 years of experience driving digital transformation across the manufacturing, energy, and industrial sectors, Matthew has led the adoption of cutting-edge solutions such as AI, robotics, IoT, and predictive analytics for Fortune 500 companies and high-growth startups alike. He has held leadership roles at organizations like Georgia-Pacific and Southern Company, where his work has contributed to major advancements in operational efficiency, safety, and sustainability.

Matthew holds a PhD in industrial engineering from the University of Tennessee, an MBA in finance from the University of Houston, and a BS in mechanical engineering from Georgia Tech. His research has been published in leading journals and has focused on machine learning applications in manufacturing and real-time monitoring systems.

He lives in Atlanta, Georgia, with his wife, their three daughters, and a "wise" 16-year-old dog. When he's not designing the future of industry, he's spending time outdoors with his family, mentoring the next generation of innovators, or sharing insights on responsible AI leadership.

Acknowledgments

To my wife and daughters—thank you for your unwavering belief in me and for your patience through every endeavor and work trip. You are my greatest motivation.

I'm grateful to my friends, colleagues, and acquaintances that I have made over the last 20 years; through those experiences and exposures, you have allowed me to explore ideas that became the foundation of this book.

Thank you Michael Carroll for your insights on the world and everything in it—our conversations over the last decade or so have inspired many of these pages.

A special thanks to Wiley and their great team of editors: without you, this would not have been possible.

Finally, thank you to the leaders, engineers, and innovators who I hope never stop striving to build a better future. This book is for you; I hope it is impactful for you.

Contents

Foreword

Occasionally, among the thousands of books written by those who think but haven't done, a book comes along from someone who has *done* and also *thinks*. This book goes beyond simply informing—it sharpens our understanding of the shiny things that capture our attention but leave us wondering how to apply their features, why they matter, or where they fit in the bigger picture. This book challenges us to rethink what's possible, offering the perspective we have—and the one we'll need. This is one of those rare books.

I've had the privilege of knowing **Matthew Alberts** for years—not just as a leader who gets things done, but as a relentless and thoughtful innovator. Matthew has a rare ability to balance action with reflection, driving change not through force but by fostering genuine connection and ownership. He deeply understands that lasting transformation doesn't come from aligning with a business case alone but from helping people feel personally invested in the outcomes.

Matthew's humility reflects his commitment to listening with curiosity rather than assuming he knows the answers—or even the right questions. His willingness to admit that he doesn't have all the answers is precisely what makes his work so powerful. He engages with teams and ideas in a way that invites participation, challenges assumptions, and creates momentum for meaningful change. His belief that progress comes from collective effort—discussing, refining, and testing ideas—is evident throughout this book.

In this timely work on **Generative AI**, Matthew brings that same clarity, vision, and humility to a subject that is already reshaping industries and human interactions. We are entering a world where the boundaries between human capability and machine intelligence are dissolving. The 6 degrees of separation that once defined our interconnectedness are collapsing into a 1-degree world—a world where people, systems, and ideas are directly and continuously linked. But connection alone isn't enough. It's what we *do* with that connection that matters.

Generative AI is the missing link between potential and realization. It isn't just a tool; it's the translator, the intermediary, that enables humans and machines to collaborate with purpose and intent. It holds the promise of shifting us from reacting to proactively creating, where intelligence isn't confined to human minds or static systems but is distributed, adaptive, and relentlessly focused on outcomes.

Matthew understands this better than most. He knows that the future doesn't belong to those who simply adopt technology—it belongs to those who demand more from it: those who insist that systems must not only automate but *reason*, that they must not only process but *understand*, and that they must not only predict but *create*. This book delivers that challenge with precision and authority.

What makes this work truly a must is not just its depth of analysis—it's the urgency of its message. Matthew compels us to look beyond the glossy marketing of AI and confront a hard truth: the difference between technology that merely serves and technology that genuinely empowers. He gives us the tools to distinguish between automation that imitates intelligence and systems you will need that *possess* it.

This is more than a book about generative AI. It's a **blueprint** for creating the end of the beginning of your new future where humans and machines collaborate not out of convenience but out of necessity—a future where intelligence is not centralized but fluid, accessible, and purpose-driven.

As you turn these pages, prepare to have your assumptions challenged. Prepare to think differently about what AI is and what it must become. But more than anything, prepare to act. Because if there's one truth this book makes clear, it's that the future isn't waiting.

Matthew has done exactly what he always sets out to do: expand the frontier of thought into action. This book is a product of that mindset, and we are all better for it.

Michael Carroll
Global AI Executive and Innovator
Retired

Introduction: A Leadership Playbook for Gen AI

Artificial intelligence is no longer a futuristic concept—it is here, transforming industries, reshaping the workforce, and redefining how organizations operate. Generative AI (Gen AI) is not just another technological advancement; it is a disruptive force that will separate companies that adapt and innovate from those that struggle to keep up. The question is no longer *if* AI will impact your business but *how* you will harness its power to drive competitive advantage.

This book is a leadership playbook for navigating the complexities of Gen AI. It provides practical strategies to help leaders

- **Build an AI-ready workforce** that collaborates seamlessly with machines
- **Integrate AI technologies** in ways that align with organizational goals and drive measurable impact
- **Address ethical dilemmas and governance challenges** to ensure responsible AI deployment

Unlike books that overwhelm you with technical jargon or abstract theories, this playbook is designed to be clear, actionable, and focused on execution. It outlines a step-by-step framework for implementing, scaling, and governing AI systems, ensuring that organizations not only adopt AI but use it strategically and responsibly.

Whether you are leading a startup, a midsized company, or a global enterprise, this book will equip you with the tools, insights, and frameworks needed to lead in the age of Gen AI. The future is not years away—it is unfolding now. We are moving from a world of intermediaries and inefficiencies to a hyper-connected, 1-degree world where AI eliminates barriers and empowers businesses like never before.

Welcome to the Gen AI revolution. Your organization's AI future starts now—let's get to work.

Generative and Agentic AI and the Dawn of the 1-Degree World

In an era where technological advancement dictates the pace of success, **Gen AI** stands as a revolutionary force reshaping industries, societies, and human interactions. The term "Gen AI" encapsulates two complementary yet distinct forms of artificial intelligence: **Generative AI** and **Agentic AI.** Together, they form the foundation of a new paradigm, enabling organizations to thrive in a world of unprecedented connectivity.

What Is Gen AI?

Gen AI is not one technology but a suite of capabilities that combines creativity with autonomy. By understanding its two dimensions—Generative AI and Agentic AI—leaders can unlock its full potential.

Generative AI: The Creative Powerhouse

AI as a concept started around 1950, with Generative AI appearing in the early 2020s, as shown in Figure 1-1. Generative AI refers to systems capable of producing new content, from writing and designing to coding and creating. Tools like GPT-4 and OpenAI generate human-like text, and DALL·E can produce intricate visuals. Beyond content creation, Generative AI fuels innovation by rapidly generating ideas that would take humans days, if not weeks. For example:

- In marketing, AI can create thousands of personalized email campaigns tailored to individual customer preferences.

- In architecture, AI tools like Midjourney and Stable Diffusion enable designers to visualize concepts instantly, cutting weeks from the design process.

- In entertainment, AI-generated scripts, animations, and music push the boundaries of creativity.

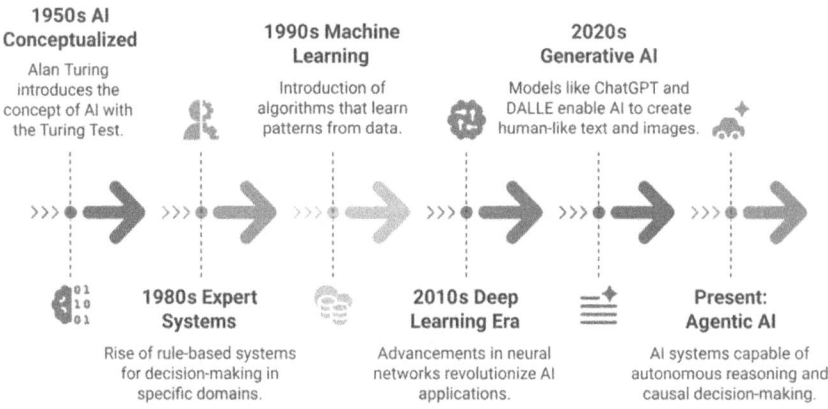

Figure 1-1: Brief history of the evolution of AI

But Generative AI doesn't stop at creation—it inspires. It offers a starting point, giving humans the freedom to refine, adapt, and innovate further.

Agentic AI: The Autonomous Problem Solver

Agentic AI, by contrast, focuses on action. These systems can reason, decide, and execute autonomously. For example:

- A supply chain AI that detects disruptions, identifies alternative suppliers, and reconfigures logistics without human intervention

- Customer service bots that not only answer queries but also resolve billing disputes or coordinate product returns seamlessly
- Autonomous vehicles that interpret real-time data to navigate roads, avoid hazards, and optimize routes

Agentic AI is not just about speed; it's about precision and adaptability. It complements Generative AI by taking creative ideas and turning them into actionable strategies.

The Synergy of Generative and Agentic AI

These two dimensions are not standalone—they thrive together. Imagine a fashion company using Generative AI to design new clothing lines based on current trends, while Agentic AI predicts customer demand, manages production schedules, and optimizes distribution. Together, they streamline creativity and execution, redefining what's possible.

The 1-Degree World: A New Era of Connectivity

The **1-degree world** is the natural outcome of evolution. In this hyper-connected reality, the distance between people, systems, and solutions shrinks to almost nothing. The traditional reliance on intermediaries—retailers, brokers, and even governments—is eroded by AI's ability to forge direct connections.

Historically, connections between individuals and resources required intermediaries. As seen in Figure 1-2, as we move toward a 1-degree world, AI and services become more distributed and better connected. For example:

- Retailers mediated the relationship between consumers and manufacturers.
- Media organizations decided what news reached the public.
- Governments and institutions acted as gatekeepers of knowledge and services.

In the 1-degree world, AI dissolves these barriers. Agentic AI facilitates direct connections, empowering individuals and businesses alike:

- **Consumers** interact directly with manufacturers for personalized products.

- **Healthcare systems** provide patients with real-time access to specialists and diagnostics without bureaucratic delays.

- **Education** becomes a one-to-one experience, with AI tutors delivering personalized learning paths.

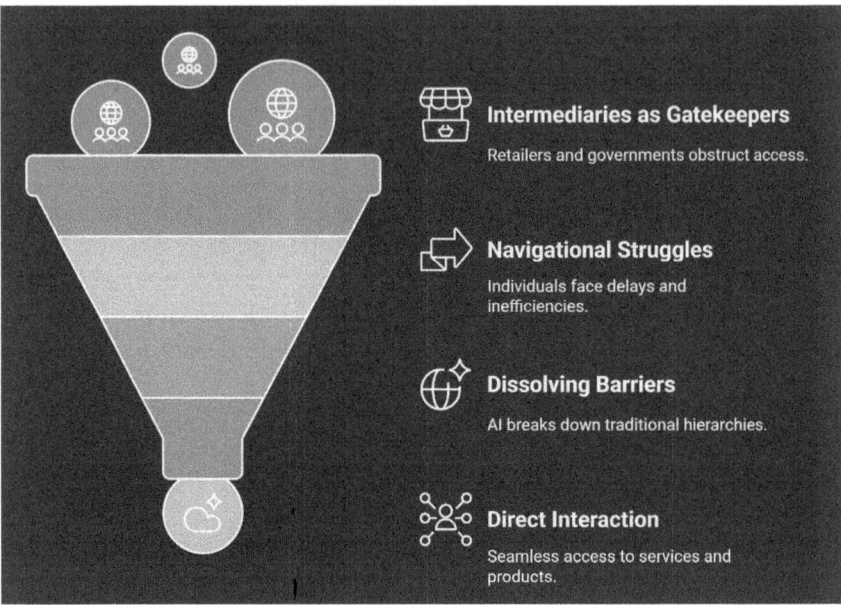

Figure 1-2: The transition from 6 degrees to 1 degree

This shift is both liberating and disruptive. For leaders, it's an opportunity to innovate, but it also requires a complete rethink of traditional power structures.

There are challenges to this hyper-connectivity. Although the 1-degree world offers unprecedented access, it also raises critical questions:

- **Equity and access:** Does everyone have equal access to AI's benefits, or do disparities in infrastructure and education create new divides?

- **Ethical concerns:** How do we prevent misuse of AI in a world without intermediaries to filter or control information?

- **Adaptation:** How can organizations built on traditional hierarchies evolve to thrive in this decentralized reality?

Why Gen AI Matters

The transformative power of Gen AI lies in its ability to amplify human capabilities. Generative AI enables creativity and scalability, and Agentic AI delivers speed and precision in decision-making that has been unseen in previous iterations of AI. Together, they allow businesses to

- **Innovate faster:** Generative AI accelerates idea generation, enabling teams to iterate rapidly. For instance, product developers can prototype multiple concepts simultaneously, reducing time-to-market.

- **Enhance efficiency:** Agentic AI reduces operational friction by **minimizing human intervention** in decision-making processes while **optimizing efficiency, responsiveness, and adaptability.** In logistics, AI can autonomously reroute shipments, saving both time and money during disruptions.

- **Personalize experiences: In manufacturing,** Gen AI transforms customer journeys by enabling companies to deliver **personalized products, services, and experiences** across the entire lifecycle—from product design to post-sale support. Whereas streaming platforms like Netflix curate content based on viewing habits, **manufacturers can apply similar principles to customize offerings, optimize production, and enhance customer satisfaction.**

For leaders, adopting Gen AI isn't optional—it's essential. The organizations that leverage AI effectively will dominate their industries, but those that hesitate risk obsolescence.

AI's Impact on Society and the 1-Degree World

The world is shrinking—not in physical space but in the degrees of separation between individuals, organizations, and systems. In this hyper-connected environment, where AI facilitates nearly direct relationships, organizations that fail to embrace this shift risk being left behind. Generative and Agentic AI (Gen AI) is not just another wave of technology; it's the foundation of a 1-degree world, where barriers are replaced by seamless translation, decision-making, and causality.

At the core of this transformation lies speed: the ability to act on insights immediately and accurately. In a world of growing complexity, humans cannot be in the middle of the decision-making process. Instead, they must remain in the loop, providing oversight and ethical judgment while AI systems handle complexity and deliver actionable recommendations in real time.

This chapter examines how the 1-degree world reshapes work and society, why causal relationships and speed are indispensable for Gen AI, and the ethical imperatives leaders must uphold as they navigate this profound shift.

The 1-Degree World: A Paradigm Shift in Connectivity

We are entering a new era of connectivity—the 1-degree world—where distance disappears, barriers dissolve, and decisions happen in real time. Enabled by advanced AI, this world allows individuals, systems, and organizations to connect directly, eliminating layers of inefficiency and unlocking unprecedented speed.

In the 1-degree world, we shift from gatekeeping to direct access. Traditional intermediaries—like retailers, brokers, or media outlets—lose their centrality as AI empowers businesses and individuals to interact with information, resources, and each other without friction.

Imagine a global manufacturer that once needed weeks to source raw materials through third-party brokers. Now, an AI-driven supply chain system connects it directly with suppliers, optimizing cost, availability, and delivery schedules in seconds—no emails, phone calls, or delays.

The key to thriving in this world is speed. Human-led processes, no matter how efficient, cannot compete with AI's ability to detect shifts, evaluate options, and initiate responses in milliseconds. The businesses that embrace AI-driven, direct connections will outpace competitors, adapt to disruptions effortlessly, and deliver the personalized, real-time experiences customers expect.

In the 1-degree world, success belongs to those who move faster than change itself.

How the 1-Degree World Redefines Decision-Making

The 1-degree world isn't just about connectivity—it's about faster, smarter, and more precise decision-making on the factory floor and across the entire supply chain. AI systems now allow manufacturers to anticipate issues, simulate solutions, and implement interventions almost instantaneously.

Here's how AI is transforming manufacturing decision-making in real time:

- **Production optimization:** Predictive AI models analyze machine performance data to detect potential breakdowns weeks before failure occurs. By automatically scheduling predictive maintenance, manufacturers reduce unplanned downtime and maintain peak production efficiency.

- **Supply chain resilience:** Gen AI connects manufacturers directly with raw material suppliers, continuously monitoring global logistics data to anticipate disruptions—like port delays or material shortages—and autonomously reroute shipments. The result? On-time deliveries, lower costs, and fewer production bottlenecks.

- **Inventory precision:** AI-driven systems dynamically adjust inventory levels based on real-time demand forecasts, machine utilization rates, and supplier reliability. This allows manufacturers to minimize excess stock while ensuring just-in-time production.

- **Quality control at scale:** Computer vision models inspect every product on the assembly line with micron-level accuracy, identifying defects or process deviations immediately. This proactive approach ensures consistent product quality while reducing rework and scrap.

In manufacturing, time is measured in margins. The companies that harness AI-driven, real-time insights to make decisions faster than competitors will achieve higher yields, lower costs, and stronger customer relationships.

In the 1-degree world, manual intervention becomes the bottleneck. AI eliminates the layers of human-mediated processes, allowing operations to run at the speed of data—with human expertise guiding the strategic decisions that matter most.

The Role of Humans

As AI systems handle increasingly complex decisions, the role of humans must evolve. Humans should not be in the middle of decision-making, where they can create bottlenecks. Rather, they should be in the loop, ensuring that decisions align with ethical principles and strategic goals.

There are reasons why humans can't be in the middle. These include complexity overload and error reduction.

In regard to complexity overload, modern data systems generate insights far faster than humans can process them. A human-mediated approach delays decision-making, leading to missed opportunities or operational inefficiencies. For example, a financial institution relying on humans to approve every trade misses critical windows of opportunity, whereas AI-driven trading platforms execute decisions in milliseconds.

With error reduction, humans are prone to cognitive biases and fatigue, which can impair decision-making. AI, when properly designed, eliminates these inconsistencies, delivering objective and reliable outcomes.

Although there are reasons humans can't be in the loop, it is also true that humans must be in the loop. These include having **ethical oversight** and **accountability**. AI systems may lack the nuanced judgment needed for ethical dilemmas. Humans provide critical oversight, ensuring that decisions align with organizational values and societal expectations. For example, in healthcare, AI may recommend aggressive treatments based on data, but a doctor must weigh the patient's preferences and overall well-being before acting.

AI systems execute decisions, but humans remain accountable for the outcomes. Ensuring a feedback loop between humans and AI allows organizations to maintain control and trust.

Causal Methodology

Causality matters: it is the key to accuracy and speed. The true power of Gen AI lies in its ability to move beyond correlation and uncover **causal relationships**—the "why" behind events. Causal methodologies enable AI systems to simulate interventions, predict outcomes, and validate decisions with unprecedented precision, allowing organizations to act faster and with greater confidence.

It is important to understand the differences between correlation and causation. Correlation identifies patterns, and causation explains them. For example:

- **Correlation**: "Customers who buy product X are likely to buy product Y."
- **Causation**: "Customers buy product Y because it complements product X in their workflows."

Additionally, understanding correlation and causation can improve decision accuracy. AI systems designed with causal methodologies consistently achieve decision accuracy rates above 95%, significantly reducing errors and enhancing trust in AI-driven insights.

This evolution in decision-making ability is an evolution not of man but rather of AI, as depicted in Figure 2-1. This evolution is made possible not just by the technology but also by the effective application of AI in business: specifically, for this book, for manufacturing operations. As we move toward a 1-degree world and the rise of Agentic AI, the ability to move faster than a competitor will be very apparent.

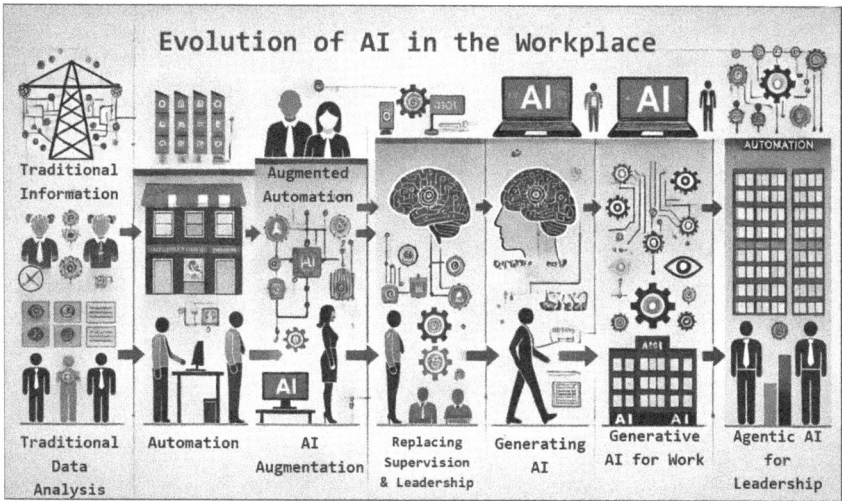

Figure 2-1: Evolution of AI in the workplace

Causal Agents in Decision-Making

To thrive in the 1-degree world, organizations must adopt **causal agents**. A causal agent is an AI system capable of doing the following:

- Identifying causal factors within complex datasets
- Testing hypothetical interventions to predict likely outcomes
- Providing actionable, evidence-based recommendations in seconds

As an example, a logistics company uses a causal agent to optimize delivery routes. By understanding the causal factors behind delays—such as traffic patterns, weather conditions, and order volumes—the AI adjusts routes in real time, reducing delivery times by 20% without human intervention.

Ethical Considerations in the 1-Degree World

A number of considerations are needed when working within a 1-degree world. These include the following:

- **Transparency and explainability:** AI systems must provide clear reasoning for their decisions to build trust and accountability. **Actionable strategy:** Use explainable AI (XAI) models that allow employees and customers to understand how decisions are made. For instance, in hiring, an AI system should explain why one candidate was selected over another.

- **Equitable access:** The 1-degree world should democratize opportunities, not deepen inequalities. Organizations must ensure that AI benefits reach all stakeholders, not just the privileged few. **Actionable strategy:** Partner with public institutions to offer AI-driven tools for underserved communities. For example, an education company could provide AI tutoring systems to schools in low-income areas.

- **Responsibility for AI outcomes:** Even as AI handles complexity, humans must retain ultimate accountability for decisions. **Actionable strategy:** Establish clear governance structures that define human oversight responsibilities, particularly in high-stakes scenarios like healthcare or finance.

Table 2-1 outlines the expected ethical challenges a strategy will face, with the appropriate mitigation strategy to reduce risk.

Table 2-1: Ethical Challenges and Mitigation Strategies

ETHICAL CHALLENGE	DESCRIPTION	MITIGATION STRATEGY
Bias and fairness	AI models can inherit biases from training data, leading to unfair outcomes.	Implement bias audits, diverse datasets, and fairness-aware AI models.
Privacy and data security	AI systems collect and process vast amounts of personal data, raising privacy concerns.	Adopt robust encryption, comply with data protection laws, and use anonymization techniques.

ETHICAL CHALLENGE	DESCRIPTION	MITIGATION STRATEGY
Misinformation and deepfakes	Generative AI can create false information, making it difficult to verify truth.	Develop AI content authentication systems, and promote digital literacy.
Job displacement	Automation can replace jobs, leading to workforce disruption and economic inequality.	Invest in reskilling programs, and create policies for AI–human workforce integration.
Lack of transparency	AI decision-making can be opaque, making it hard to explain why certain outcomes occur.	Ensure AI explainability through interpretable models and regulatory frameworks.
Accountability and governance	Unclear responsibility for AI-driven decisions can lead to legal and ethical issues.	Establish clear governance policies, AI ethics boards, and accountability measures.

Leadership in the 1-Degree World

Leaders play a pivotal role in driving AI adoption while maintaining ethical and strategic oversight. Key leadership actions include

- **Fostering speed-driven cultures**: Encourage employees to embrace AI tools that enhance decision-making speed without compromising accuracy. Leadership must champion the idea that agility is critical to success.

- **Investing in causal methodologies**: Prioritize AI systems designed for causal reasoning to unlock actionable insights that align with organizational goals. Leaders should allocate resources to integrate these systems across key functions.

- **Promoting AI literacy**: Equip employees with the knowledge to trust and utilize AI effectively, ensuring seamless integration into workflows. Offer training programs that highlight the importance of being in the loop without becoming bottlenecks.

Embracing the 1-Degree World

The 1-degree world is not just a technological shift; it's a paradigm change in how decisions are made and actions are taken. By leveraging speed, causal methodologies, and the ability to translate insights into action, organizations can unlock the true power of Gen AI. Leaders must guide this transformation thoughtfully, ensuring that humans remain in the loop to provide oversight and maintain ethical alignment.

Organizations that fail to adapt will be left behind, trapped in outdated processes that cannot match the speed or accuracy of AI-driven competitors. The future belongs to those who embrace the 1-degree world—not as a challenge but as a profound opportunity.

Key Business Functions Transformed by Gen AI

Gen AI is reshaping businesses across the board, fundamentally transforming how core functions operate. From marketing and sales to product development, operations, and human resources, AI is creating new efficiencies, driving growth, and enabling deeper personalization. Understanding how these changes impact different business functions is crucial for leaders looking to implement AI solutions effectively. Table 3-1 provides an overview of how AI is revolutionizing key business functions, highlighting traditional versus AI-enhanced approaches and their benefits.

Table 3-1: AI's Impact on Key Business Functions

BUSINESS FUNCTION	TRADITIONAL APPROACH	GEN AI-ENHANCED APPROACH	KEY BENEFITS
Marketing and sales	Manual customer segmentation, generic email campaigns	AI-driven personalized content, dynamic customer insights	Higher engagement, increased conversions, efficiency

Continues

Table 3-1 (*continued*)

BUSINESS FUNCTION	TRADITIONAL APPROACH	GEN AI-ENHANCED APPROACH	KEY BENEFITS
Product development	Manual R&D, slow iteration cycles	AI-generated design suggestions, virtual prototyping	Faster innovation, reduced costs, higher customer satisfaction
Operations and supply chain	Scheduled maintenance, reactive planning	AI-powered predictive analytics for inventory and logistics	Reduced downtime, cost savings, improved efficiency
Customer service	Call centers, long wait times	AI chatbots, automated support, sentiment analysis	24/7 support, faster response, improved customer satisfaction
HR and talent management	Manual hiring, generic training programs	AI-driven resume screening, personalized learning paths	Faster hiring, better employee retention, enhanced productivity

The comparison shown in Table 3-1 helps business leaders understand how AI can be leveraged across various departments. As AI adoption grows, companies that integrate AI effectively will gain a competitive edge by increasing efficiency, reducing costs, and improving customer satisfaction.

As we move through the remainder of this chapter, we will explore how Gen AI is transforming core business functions across industries. We'll delve into the ways AI-driven personalization is reshaping **marketing and sales**, driving customer engagement and increasing conversion rates. Next, we'll examine how AI accelerates **product development** by enhancing design cycles, optimizing resources, and driving innovation. We'll also look at how **operations** benefit from AI's predictive capabilities, streamlining supply chains and improving process efficiency. Finally, we'll discuss how **human resources** is leveraging AI to improve talent acquisition, personalize employee development, and foster a more adaptive, high-performing workforce. Through these examples and case studies, we'll illustrate how businesses can strategically deploy Gen AI across functions to unlock new levels of productivity and growth.

Marketing and Sales: Personalization and Content Creation at Scale

One of the most visible impacts of Gen AI is in marketing and sales. AI's ability to generate personalized, dynamic content allows companies to engage with customers on a much deeper level. Traditionally, creating personalized marketing campaigns required significant time and resources. With AI, businesses can generate thousands of personalized emails, advertisements, and social media posts in minutes, each tailored to the unique preferences and behaviors of individual customers. Figure 3-1 illustrates how AI enhances marketing personalization through a step-by-step process.

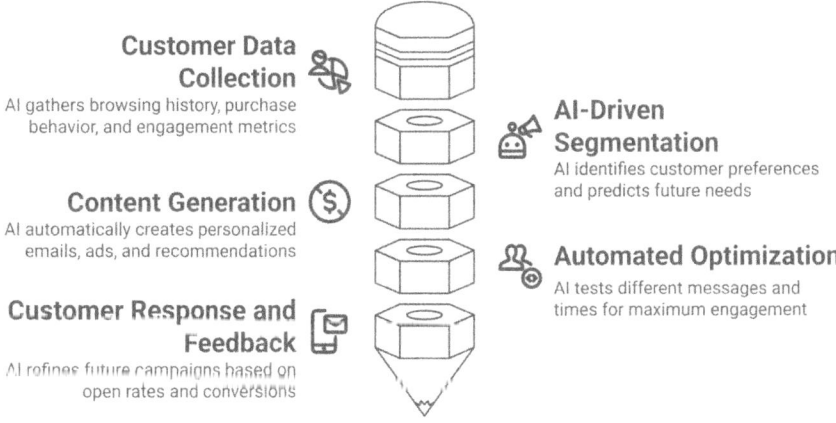

Figure 3-1: AI-powered marketing personalization flowchart

The structured approach shown in Figure 3-1 allows companies to scale personalized marketing efforts without significantly increasing costs or human workload. By leveraging AI, businesses can ensure the right message reaches the right customer at the right time, significantly boosting engagement and conversion rates.

Gen AI also enables more precise customer segmentation. By analyzing vast amounts of data, AI can predict customer behavior, identify trends, and suggest personalized products or services in real time. This allows sales teams to focus on higher-value interactions while AI handles the repetitive but crucial tasks of lead nurturing and customer follow-up.

Let's look at an example along with the challenge it brought, the solution that was implemented, and the results attained.

Example: AI-Driven Personalized Marketing for a Medium-Sized E-Commerce Company   A medium-sized e-commerce company specializing in fashion and home goods faced challenges in delivering personalized experiences without expanding its marketing team. Its generic email campaigns led to low engagement and limited conversions. The company needed a solution that could automate personalized content creation at scale without requiring significant manual effort.

Challenge   With a small marketing team, the company struggled to create tailored campaigns for its growing customer base. Segmenting customers manually based on their purchase histories and preferences was time-consuming, leading to one-size-fits-all emails that lacked relevance.

Solution   The company implemented an AI-driven marketing solution to address this. Using machine learning algorithms, the AI analyzed customer data, including purchase histories, browsing behaviors, demographic details, and engagement patterns. The system automatically segmented customers and generated personalized email content. For example:

- Customers who frequently purchased shoes received recommendations for matching accessories.

- Those browsing summer collections were targeted with discounts on similar items.

Each email was customized with dynamic content, and the AI system optimized sending times based on when recipients were most likely to engage. It also predicted future purchases, allowing the company to offer relevant product recommendations.

Results   Over six months, the AI-enhanced strategy led to a 35% increase in email open rates and a 25% boost in conversions. Emails resonated more with customers due to the personalized recommendations, resulting in greater engagement. The sales pipeline grew substantially, driven by the AI's ability to identify and convert high-value customers, increasing both the average order value (AOV) and customer lifetime value (CLV).

Collaboration Between AI and Human Teams   The marketing team focused on strategy while the AI handled segmentation and content generation, enabling the company to scale its efforts without additional hires.

Cost Savings and Efficiency   The AI allowed the company to scale its marketing efforts without hiring additional staff. Tasks like data analysis, segmentation, and content creation were automated, freeing the marketing team to focus on higher-level strategy. This not only saved costs but also significantly improved operational efficiency.

Scalability and Strategic Impact   The AI system continuously improved over time, leading to even more precise targeting in future campaigns. Insights from the AI's analysis helped refine broader marketing strategies, including promotions and inventory management. The success of the marketing AI led the company to explore AI applications in other areas, marking the beginning of a broader digital transformation.

Conclusion   The AI-driven platform streamlined marketing operations, boosted customer engagement, and demonstrated the potential for AI to personalize experiences at scale.

Product Development: Accelerating Innovation Cycles

Gen AI is playing a pivotal role in speeding up product development cycles. Companies can now leverage AI models to generate new product ideas, designs, and even fully functional prototypes. This is particularly useful for businesses that need to stay ahead of market trends and release products faster than their competitors. For businesses that rely on rapid innovation, AI drastically reduces the time required for design, prototyping, and iteration. Traditionally, product development cycles could take months or even years due to lengthy manual processes. The timeline shown in Figure 3-2 compares traditional product development with AI-driven acceleration.

By integrating AI, companies can reduce the time spent on repetitive tasks and focus on refining and perfecting the most promising product designs. By analyzing customer feedback, market data, and competitor products, AI systems can propose new designs or features that align with current consumer demands. This not only reduces the time spent on research and development (R&D) but also helps businesses mitigate the risk of launching products that fail to meet market expectations. Let's look again at an example.

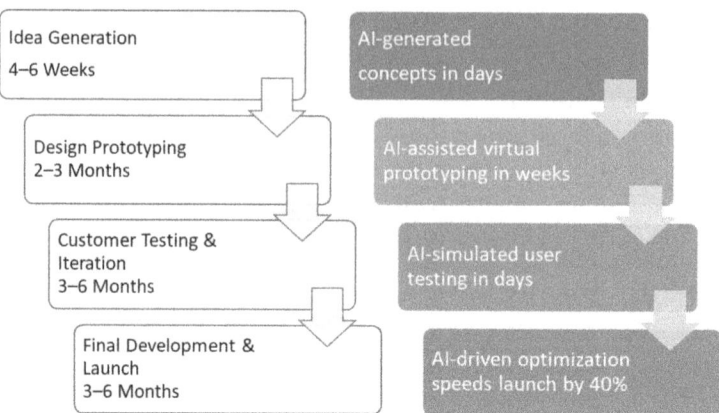

Figure 3-2: Product development acceleration timeline

Example: Accelerating Product Development with Gen AI at a Global Technology Company   A global technology company known for its consumer electronics faced increasing pressure to release new products faster. Traditional product development, which relied on manual design iterations and prototyping, was slowing the company down, leading to delayed product launches and missed market opportunities. To address this, the company integrated Gen AI into its product development pipeline.

Challenge   The company's development cycle involved multiple stages of manual design, customer feedback analysis, and prototyping. This process was too slow to keep pace with competitors, limiting the company's ability to innovate quickly.

Solution   The company implemented Gen AI, which analyzed customer feedback, market trends, and product data to generate multiple design concepts. The AI

- **Analyzed customer feedback** from previous products to identify patterns and preferences
- **Identified market trends** by monitoring competitor products
- **Generated design concepts** optimized for functionality, aesthetics, and cost
- **Ran simulations** to test designs virtually, reducing the need for physical prototypes

For example, in developing a new wearable device, the AI suggested lightweight, flexible materials that addressed customer concerns about comfort while maintaining the product's premium feel.

Results   Gen AI reduced the company's product development time by 40%. It allowed designers to explore and evaluate hundreds of design variations in parallel, eliminating low-performing ideas early. The AI-driven insights from customer feedback also improved product alignment with market demands, leading to a 30% increase in customer satisfaction for new releases. For instance, a new smartphone model featured AI-driven battery optimization, which directly addressed prior customer complaints and led to significantly improved reviews.

Collaboration Between AI and Human Teams   Engineers reviewed and refined AI-generated designs, accelerating the innovation process.

Cost Savings and Efficiency Gains   The virtual prototyping process significantly cut costs associated with physical models.

Scalability and Strategic Impact   The company integrated AI into supply chain forecasting and inventory management.

Conclusion   AI-enhanced product development reduced time-to-market, improved customer satisfaction, and positioned the company as an innovation leader.

Operations: Enhancing Efficiency with Predictive Analytics

Operations management has traditionally relied on historical data and manual processes to optimize performance. With Gen AI, companies can use predictive analytics to streamline operations, forecast demand, and manage supply chains more effectively. AI models can analyze patterns in production data, detect inefficiencies, and provide real-time recommendations for improving workflows.

AI also enhances supply chain management by predicting disruptions and optimizing inventory levels based on real-time data. This leads to more efficient resource allocation and reduces waste, saving companies both time and money.

Example: AI-Powered Predictive Maintenance at a Large Manufacturing Firm   A large manufacturing firm operating multiple global production facilities faced significant challenges with maintaining its machinery. Traditional maintenance practices—based on scheduled inspections or reactive repairs—led to frequent breakdowns, costly downtime, and inefficient resource use.

The firm needed a more proactive solution to monitor and maintain its equipment effectively.

Challenge   The firm's preventive maintenance approach was inefficient and reactive, often leading to unexpected machine failures that caused production delays and high repair costs. Scheduled maintenance also wasted resources by servicing machines unnecessarily, and critical breakdowns still disrupted operations.

Solution   The company implemented an AI-powered predictive maintenance system, which used data from sensors on machinery to analyze performance in real time. The AI system continuously monitored factors like temperature, vibration, and pressure, predicting potential failures before they happened. This allowed the company to schedule maintenance only when necessary, shifting from a time-based to a condition-based strategy.

Key features of the AI system included

- **Real-time data collection** from factory floor sensors
- **Predictive analytics** that identified early warning signs of machine failure
- **Automated alerts** to notify maintenance teams of impending issues, allowing for preemptive repairs
- **Optimized maintenance schedules** based on actual machine conditions rather than arbitrary timelines

Results   The following results were attained:

- **60% reduction in downtime**: By predicting failures, the AI system allowed the company to prevent unexpected breakdowns, scheduling maintenance during planned downtimes and minimizing production disruptions.
- **Millions in cost savings**: The ability to prevent costly emergency repairs and reduce unnecessary part replacements saved the company millions. AI also helped the firm avoid over-maintenance, cutting down on wasted labor and resources.
- **25% improvement in operational efficiency**: With fewer breakdowns and more efficient maintenance, the company achieved smoother production flows and increased throughput across its factories.

Collaboration Between AI and Human Teams   AI handled complex data analysis and predictions, and maintenance teams used these insights to prioritize tasks and address the most critical issues first. This human–AI collaboration resulted in faster, more informed maintenance decisions and improved overall equipment reliability.

Cost Savings and Efficiency Gains   Preventing breakdowns and optimizing maintenance schedules lowered costs by 60%.

Scalability and Strategic Impact   Following its initial success, the firm scaled the AI system to other global production sites. The system seamlessly integrated with the company's existing Internet of Things (IoT) infrastructure and machinery. In the long term, this predictive maintenance strategy contributed to sustainability goals by reducing energy use and extending machinery lifespans. It also paved the way for future AI applications in production optimization and quality control.

Conclusion   The AI-powered predictive maintenance system transformed the firm's approach to machinery upkeep, enabling a proactive strategy that reduced downtime by 60%, saved millions in repair costs, and improved operational efficiency by 25%. The integration of AI not only optimized production processes but also empowered human teams with data-driven insights for better decision-making.

Customer Service: Transforming Customer Interactions with AI

Customer service is undergoing a significant transformation with the rise of AI-powered chatbots and virtual assistants. These tools, often powered by natural language processing (NLP), can handle routine customer inquiries, process orders, and provide 24/7 support without the need for human intervention. For businesses with large customer bases, AI ensures that customers receive timely responses and consistent service quality, regardless of the time of day.

AI ensures 24/7, high-quality customer service through several key mechanisms:

- **Natural language processing (NLP):** AI systems use NLP to understand, interpret, and respond to customer inquiries in a human-like manner. This allows chatbots and virtual assistants to handle routine questions, process orders, and provide personalized assistance without human intervention.

- **Automated query handling:** AI models are trained on vast datasets of customer interactions, enabling them to quickly recognize common queries and provide accurate responses. Frequently asked questions, like order status, returns, and product information, can be handled automatically, reducing wait times and ensuring consistency.

- **Real-time data access:** AI-driven systems integrate with internal databases—such as CRM, inventory, and billing systems—to access real-time information. For example, a chatbot can instantly retrieve an order's shipping status or product availability without human assistance.

- **Continuous learning and adaptation:** Machine learning algorithms continuously analyze interactions to improve response accuracy over time. If a particular query type becomes more common, or if customers react negatively to certain responses, the system adapts its approach accordingly.

- **Load balancing and prioritization:** AI monitors query volumes and prioritizes interactions based on urgency and complexity. Critical or time-sensitive issues can be escalated to human agents, ensuring that high-priority requests receive immediate attention and routine questions are resolved autonomously.

- **Sentiment analysis:** Advanced AI tools apply sentiment analysis to detect customer emotions during interactions. If a customer shows signs of frustration, the system can escalate the issue to a human agent, ensuring a better experience.

- **Predictive assistance:** AI can anticipate customer needs based on past interactions. For instance, if a customer recently purchased a product, the chatbot might proactively offer setup instructions or troubleshooting tips.

By combining these capabilities, AI-powered systems ensure fast, accurate, and reliable customer service around the clock—improving customer satisfaction while freeing human agents to handle more complex or nuanced issues.

Example: AI-Driven Customer Service Transformation at a Telecommunications Company   A large telecommunications company faced challenges in handling high volumes of customer inquiries, which led to long wait times and rising customer dissatisfaction. The company sought to improve response times and reduce operational costs while maintaining service quality.

Challenge   The company's call centers were overwhelmed with repetitive inquiries, leading to slow response times and high operational costs. It needed a way to automate routine tasks while still providing excellent customer service.

Solution   The company implemented an AI-powered chatbot across its website, mobile app, and social media platforms, designed to handle a wide range of customer inquiries autonomously. The chatbot was trained to

- **Handle common questions** (billing, technical support, service changes)
- **Perform routine tasks** (password resets, plan upgrades)
- **Escalate complex issues** to human agents when needed

In addition to managing conversations, the AI analyzed chatbot interactions to identify frequent pain points and improve customer service processes.

Results   The following results were attained:

- **70% of inquiries resolved without human intervention:** The chatbot handled most inquiries autonomously, reducing the workload of human agents and allowing them to focus on more complex issues. For example, customers could easily reset their internet connection via the chatbot.

- **50% reduction in response times:** With the AI chatbot managing the majority of inquiries, response times were halved, offering 24/7 support and quick resolutions without the need for long wait times.

- **20% increase in customer satisfaction:** By analyzing chatbot conversations, the company identified common customer frustrations, such as unclear billing. This led to process improvements, contributing to a 20% increase in satisfaction scores.

Collaboration Between AI and Human Teams   While the chatbot handled routine tasks, human agents took over complex cases, aided by the AI's conversation history. This AI–human collaboration improved efficiency and allowed agents to provide more personalized support when needed.

Cost Savings and Efficiency Gains   By automating 70% of customer inquiries, the company reduced labor costs and improved resource allocation. The chatbot's continuous learning also optimized responses and processes over time, increasing overall efficiency.

Scalability and Strategic Impact   The success of the chatbot led to its expansion into other areas like sales and marketing. With NLP enhancements, the chatbot's capabilities continued to grow, allowing it to handle more complex conversations and further reduce human intervention.

Conclusion   By deploying an AI-powered chatbot, the telecommunications company reduced customer service response times by 50%, resolved 70% of inquiries without human involvement, and improved customer satisfaction by 20%. The integration of AI transformed both operational efficiency and the customer experience, paving the way for future AI-driven innovations.

Human Resources: AI for Talent Acquisition and Employee Development

In human resources, Gen AI is transforming talent acquisition, employee training, and performance management. AI-driven recruitment tools can scan resumes, assess candidates based on predefined criteria, and even conduct initial interviews, significantly reducing the time and resources needed to find the right candidates. AI can also help reduce unconscious bias in hiring by focusing solely on the qualifications and experiences that matter.

AI is also changing how companies approach employee training. By analyzing employee performance data, AI systems can recommend personalized training programs that align with individual strengths and development needs. This ensures that employees continue to grow and remain engaged in their roles.

Example: AI-Driven Recruitment and Training at a Financial Services Company   A financial services company faced challenges with long recruitment times and varying new hire performance. To address this, the company implemented AI-powered solutions to streamline both its recruitment process and employee training programs, leading to significant improvements in efficiency and employee outcomes.

Challenge   The company's traditional recruitment process was slow and manual, often taking weeks to screen resumes and conduct initial interviews. High competition for talent made it difficult to secure candidates before they accepted other offers. Additionally, new hires often faced a steep learning curve, resulting in inconsistent performance and lower retention.

Solution   The company deployed an AI-driven recruitment platform that automated resume screening, candidate assessments, and predictive analytics to identify the best fit for each role. This reduced the manual burden on HR and sped up the hiring process.

The AI-enhanced recruitment included the following:

- Automated resume analysis filtered unqualified candidates based on preset criteria, saving time for recruiters.
- AI-driven assessments, including video interviews and technical tests, evaluated candidates on skills and cultural fit.
- Predictive analytics forecasted candidate success based on historical hiring data, improving the quality of hires.

The company also introduced AI-driven training programs for new hires, offering personalized learning paths tailored to each individual's strengths and weaknesses. This adaptive training helped new employees develop the skills they needed to perform well in their roles from day one.

The AI-powered personalized training included

- Custom learning plans that addressed skill gaps based on each hire's background
- Adaptive training that adjusted content in real time to match employees' learning progress, providing more support where needed
- AI monitoring of performance during training, allowing HR to track progress and offer additional development where necessary

Results   The following results were attained:

- **40% reduction in time-to-hire:** AI-driven automation cut the recruitment timeline by 40%, reducing it from six to four weeks. Faster hiring allowed the company to secure top talent before competitors.
- **15% improvement in new hire performance:** Personalized, AI-curated training paths helped new hires improve their performance by 15%, as employees were better prepared and more productive from the start.
- **10% increase in employee retention:** The tailored training programs led to a 10% boost in employee retention, as new hires felt more supported and engaged in their career development.

Collaboration Between AI and Human Teams   Recruiters focused on strategic tasks while AI handled administrative processes.

Cost Savings and Efficiency Gains   By automating much of the recruitment process and improving new hire readiness through tailored training, the company saved on labor costs and reduced time-to-productivity. The higher retention rates also lowered costs associated with turnover and rehiring.

Scalability and Strategic Impact   Building on the success of AI in recruitment and training, the company began using AI to create career development paths for existing employees, identify leadership potential, and help employees upskill for future roles.

Conclusion   Through AI-powered recruitment and training, the company cut its time-to-hire by 40%, improved new hire performance by 15%, and boosted retention by 10%. These AI-driven solutions created efficiencies across the board while enhancing the employee experience, positioning the company for sustained growth and competitiveness.

Revolutionizing Key Business Functions

As demonstrated throughout this chapter, Gen AI is revolutionizing key business functions, enhancing efficiency, optimizing decision-making, and driving innovation across industries. Whether in marketing and sales, product development, operations, customer service, or human resources, AI is not just an enabler—it is a force multiplier that allows businesses to scale operations, personalize experiences, and remain competitive in an increasingly digital landscape.

The real power of Gen AI lies in its ability to automate routine tasks while empowering human teams to focus on higher-value, strategic initiatives. In marketing, AI transforms customer engagement through hyper-personalized content and predictive insights. In product development, it accelerates innovation cycles by generating designs and running simulations, reducing time-to-market. AI-driven predictive analytics in operations optimize supply chains and maintenance, leading to significant cost savings. Meanwhile, AI-powered chatbots and virtual assistants revolutionize customer service, ensuring 24/7 support and improved satisfaction. Finally, in human resources, AI streamlines talent acquisition and development, fostering a more agile and skilled workforce.

As businesses continue to integrate AI into their core functions, the challenge will not be whether to adopt AI but how to do so strategically and effectively. Companies that embrace AI-driven transformation today will be better positioned to harness its full potential in the future. The next chapter will delve deeper into the role of AI strategy, offering insights into how organizations—whether small, medium, or large—can tailor AI adoption to their specific needs and maximize long-term success.

Tailoring AI Strategy for Small, Medium, and Large Organizations

As organizations consider deploying Gen AI to transform operations, one thing becomes clear: there's no one-size-fits-all approach. The strategies for adopting and scaling AI differ significantly depending on the size of the company, available resources, operational complexity, and internal capabilities. Understanding these differences helps leaders implement AI in a way that aligns with their organization's needs and strategic goals.

Businesses are generally categorized into small, medium, and large enterprises based on several factors, including workforce size, annual revenue, and operational scale:

Small businesses: Typically, small businesses have fewer than 100 employees and generate annual revenues of less than $10 million. These companies often have lean structures, limited budgets, and minimal in-house technical expertise. As a result, their AI adoption strategies prioritize low-cost, high-impact solutions that address immediate operational challenges, such as customer support automation or process optimization.

Medium-sized businesses: Medium enterprises usually employ between 100 and 1,000 people and report annual revenues ranging from $10 million to $1 billion. These organizations have more resources than small businesses but must carefully manage investments as they scale. Their AI strategies often focus on expanding capabilities, integrating AI across departments, and building a foundation for future growth while maintaining operational agility.

Large enterprises: Large companies typically have more than 1,000 employees and annual revenues exceeding $1 billion. With complex global operations and vast amounts of data, these businesses require robust AI governance frameworks, scalable infrastructure, and cross-functional collaboration to successfully implement and manage AI initiatives across different markets and departments.

By tailoring AI adoption strategies to organizational size, leaders can drive sustainable value, maximize return on investment (ROI), and ensure smooth implementation. Table 4-1 summarizes the key distinctions and considerations for AI deployment across small, medium, and large businesses, providing a roadmap for leaders to craft strategies that fit their organization's unique characteristics.

Table 4-1: Tailoring AI Adoption for Small, Medium, and Large Organizations

FACTOR	SMALL BUSINESSES	MEDIUM BUSINESSES	LARGE BUSINESSES
Budget constraints	Limited funds, need low-cost AI solutions	Moderate investment with need for clear ROI	Large AI budgets for R&D and global infrastructure
AI talent and expertise	No in-house AI experts, rely on no-code tools	May hire specialists or partner with vendors	In-house AI teams, AI Centers of Excellence
Scalability	Start small, focus on targeted AI applications	Ensure that AI solutions can scale with growth	Enterprise-wide AI deployment across departments
AI use cases	Automating tasks (chatbots, email marketing, bookkeeping)	AI-driven insights for marketing, sales forecasting, and logistics	Advanced AI: real-time predictive analytics, fraud detection, deep learning models
Governance and compliance	Minimal AI governance, focus on rapid execution	Some AI policies for risk mitigation	Strong governance, compliance with AI ethics and regulations

Table 4-1 helps business leaders understand how AI adoption varies across different organizational sizes. Small businesses should focus on high-ROI, off-the-shelf AI solutions, whereas larger enterprises must ensure that AI aligns with governance policies and long-term scalability. By understanding these differences, organizations can create a more effective AI strategy that aligns with their business goals.

This chapter explores the key considerations for deploying AI across small, medium, and large organizations, providing actionable steps for leadership in each context. I encourage you to read the considerations for all the different organizational sizes. After all, large organizations usually contain smaller and medium internal groups.

Small Organizations: Leveraging AI for Maximum Impact with Minimal Resources

For small businesses, the goal of AI adoption is often about gaining competitive advantage through efficiency without breaking the bank. With limited resources and smaller teams, small organizations typically need to be selective about where they apply AI, focusing on areas where they can achieve immediate, measurable results.

Challenges for Small Organizations   There are a number of challenges to consider for smaller organizations. These include

- **Limited budget:** Smaller firms generally have less capital to invest in cutting-edge AI tools or dedicated data science teams.

- **Lack of in-house expertise:** Small businesses often lack AI specialists or data scientists on staff, making it challenging to develop custom AI solutions.

- **Scalability concerns:** Implementing AI solutions at a small scale can be daunting, particularly when considering future growth. This can be due to a lack of human or financial resources or other factors that influence small businesses more than larger ones.

Strategic Focus   Having a strategy when focusing on the areas can include

- **Starting with off-the-shelf AI tools:** AI platforms that require minimal technical expertise are ideal for small businesses. Tools like AI-powered customer service chatbots, email marketing automation, and AI-driven data analytics can deliver significant ROI without a massive upfront investment.

- **Customer service and chatbots:**
  - Zendesk Answer Bot: AI-driven customer support that automates responses to common inquiries and escalates complex issues to human agents
  - Intercom: Provides conversational AI tools for customer engagement, including live chat, product tours, and automated support
  - Drift: An AI-powered chatbot that helps businesses engage with website visitors in real time and qualify leads without manual effort
- **Email marketing automation:**
  - Mailchimp: Integrates AI to personalize email content, optimize send times, and automate customer journeys
  - ActiveCampaign: Uses predictive analytics and machine learning to create dynamic, personalized email campaigns based on customer behavior
  - HubSpot: Includes AI-driven email marketing features such as smart content recommendations and automated workflows
- **AI-driven data analytics:**
  - MonkeyLearn: A no-code text analysis platform that uses machine learning to analyze customer feedback, social media interactions, and survey responses
  - Pecan AI: Predictive analytics tool that helps businesses forecast trends and make data-driven decisions without a data science team
  - Tableau with Einstein Discovery (Salesforce): Provides intuitive dashboards with AI-powered insights and recommendations
- **Marketing automation and content generation:**
  - Jasper AI: Generates marketing copy, blog posts, and social media content using natural language processing (NLP)
  - Canva Magic Write: Built into Canva; uses AI to generate text for social media posts, blogs, and ads
  - Copy.ai: An AI-powered copywriting tool for creating marketing content quickly and effectively

- **HR and talent management:**
  - Breezy HR: Uses AI to streamline recruitment, automate candidate screening, and manage interview scheduling
  - Zoho Recruit: AI-driven talent acquisition platform that helps identify the best candidates by analyzing resumes and applications
  - Hiretual: AI sourcing tool that identifies talent across various platforms and suggests candidates based on job requirements

For example, a local bakery used Pecan AI for inventory management tool to track ingredient usage and predict demand.

AI in action:

- Pecan AI analyzed historical sales data, social media trends, and search activity.
- The platform predicted a 30% increase in demand during different demand cycles.
- The team adjusted marketing budgets and inventory accordingly.

Outcome:

- Reduced stockouts by 30%
- Achieved a 22% lift in revenue compared to the previous year
- Reduced marketing spend by 15% by targeting the most responsive customer segments

- **Focusing on specific use cases:** Remember, we are employing a strategic focus. Instead of attempting to deploy AI across multiple functions, small businesses should focus on one or two high-impact areas. For example, using AI to improve customer interactions or automating repetitive tasks like bookkeeping can free up staff for more strategic work.

- **Leveraging cloud-based AI solutions:** Cloud platforms such as Amazon Web Services (AWS), Google Cloud AI, and Microsoft Azure AI provide small businesses with scalable, pay-as-you-go solutions that don't require significant infrastructure investment.

Expanded Example　A small digital marketing agency wanted to offer more personalized services to clients without adding to its workforce. By integrating Jasper AI, the company automated personalized email campaigns and social media interactions based on customer behavior and engagement.

AI in action:

- Jasper created SEO-friendly blog drafts and product descriptions.
- The team used AI to generate social media captions in bulk for various platforms.
- The agency trained Jasper on each client's brand voice for consistent messaging.

Outcome:

- 70% faster content production
- Increased client satisfaction by delivering content ahead of deadlines
- Scaled to support 20% more clients without adding headcount

This strategy allowed the agency to compete with larger firms while staying within budget.

Medium-Sized Organizations: Balancing Growth and Innovation with AI

Medium-sized businesses generally have more resources than small companies but face unique challenges when scaling AI. Given that they fall in an area where some companies may have more roles to focus on transformation initiatives and others may not, they need a balance most of all. These organizations can afford to invest in AI but need to balance that investment with the ongoing demands of growth. For medium enterprises, the challenge is finding the right balance between implementing scalable AI solutions and maintaining operational agility.

Challenges for Medium-Sized Organizations   There are a number of challenges to consider for medium-sized organizations. These include

- **Balancing cost and innovation:** Medium-sized businesses often have larger budgets but need to demonstrate measurable ROI quickly.

- **Scalability:** AI solutions implemented today must scale seamlessly as the company grows.

- **Talent gaps:** Medium-sized firms often struggle to attract and retain AI talent, making it harder to develop in-house AI solutions.

Strategic Focus   Having a strategy when focusing on the areas can include

- **Prioritizing cross-functional AI initiatives:** AI should not be siloed within one department but integrated across multiple functions like marketing, sales, and operations. Cross-functional collaboration enhances both efficiency and innovation.

- **Investing in AI talent or partnering with AI vendors:** Medium-sized organizations can hire dedicated AI professionals or partner with third-party vendors to bridge the gap between technical capability and business needs.

  For example, a regional healthcare provider serving multiple clinics and outpatient centers struggled with inefficient scheduling processes, resulting in long patient wait times, staff burnout, and missed appointment opportunities. The organization sought a solution that could optimize staff allocation and streamline appointment scheduling without disrupting daily operations.

  The provider partnered with Qventus, an AI vendor specializing in healthcare operations optimization. Qventus implemented an AI-powered predictive scheduling system to

  - Analyze historical appointment patterns, no-show trends, and patient demographics

  - Predict demand fluctuations and dynamically adjust scheduling to match staffing needs

  - Provide real-time alerts to administrators when appointment slots needed adjustment

AI in action:

- The system adjusted appointment durations based on historical trends (e.g., new patient visits took longer than follow-ups).

- Staff schedules were optimized by predicting peak demand periods.

- Automated reminders and rescheduling prompts were sent to reduce no-shows.

Outcome:

- 30% reduction in patient wait times
- 20% improvement in staff utilization, particularly for specialists
- 15% decrease in no-shows, thanks to AI-driven reminders
- Higher patient satisfaction scores, with positive feedback about more predictable appointment experiences

- **Piloting AI projects before scaling:** Medium-sized businesses are well-positioned to experiment with pilot AI programs. For instance, piloting an AI-driven sales forecasting tool in one region can reveal its potential impact before scaling across the organization.

Expanded Example A mid-sized logistics company with a fleet of 500 delivery vehicles faced escalating operational costs due to rising fuel prices, inefficient routing, and frequent delivery delays. The company operated across multiple cities but wanted to pilot a solution in one urban area to validate its effectiveness before committing to a larger rollout.

The logistics company struggled with

- Rising fuel costs due to inefficient routing and idling in traffic
- 15% late delivery rate, impacting customer satisfaction
- Lack of visibility into route bottlenecks and real-time fleet performance

Solution:

The company partnered with OptimoRoute, an AI-powered route optimization platform, to pilot a solution in Atlanta, GA—a city known for complex traffic patterns. The AI system was integrated with the company's existing telematics and dispatch systems to

- Analyze real-time traffic data, fuel consumption patterns, and delivery schedules
- Generate optimized routes that minimized fuel consumption and avoided traffic hotspots
- Provide real-time route adjustments based on traffic incidents or weather conditions

AI in action:

- The system identified previously unnoticed traffic bottlenecks near distribution centers.

- Dynamic rerouting capabilities allowed drivers to avoid congested areas.
- Predictive analytics helped optimize delivery schedules based on historical traffic patterns.

Outcome:

- 20% reduction in fuel costs, thanks to more efficient routing and reduced idle time
- 15% improvement in delivery times, with fewer missed delivery windows
- Insights into operational inefficiencies, such as loading dock delays and suboptimal driver assignments
- By starting small, the company minimized risk while gaining valuable insights for broader implementation

Large Organizations: Managing Complexity and Scaling AI Globally

Large enterprises face the challenge of managing greater complexity when deploying AI. With multiple departments, global operations, and vast amounts of data, large organizations can benefit tremendously from AI—but only if it is implemented correctly. Scaling AI across such a vast operation requires careful planning, significant investment, and a robust governance structure. Large organizations must implement AI governance frameworks to ensure ethical, transparent, and responsible AI use. These frameworks focus on key areas such as data privacy, bias mitigation, security, and accountability, aligning AI deployment with corporate and regulatory standards. A well-structured governance framework includes policies on AI ethics, outlining fairness, transparency, and responsible use.

Challenges for Large Organizations   There are a number of challenges to consider for large organizations. These include

- **Data management:** Large organizations generate vast amounts of data, which can be difficult to manage and integrate across various AI systems.
- **AI governance:** Ensuring responsible and ethical AI use is crucial, given the higher levels of scrutiny faced by large enterprises.

- **Cross-departmental coordination:** Deploying AI often requires collaboration across multiple departments, which can pose significant logistical challenges.

Strategic Focus   Having a strategy when focusing on the areas can include

- **Establishing AI governance and ethical standards:** Large organizations must implement robust governance frameworks to ensure ethical AI use, including policies on data privacy, bias mitigation, and accountability.

- **Creating AI Centers of Excellence (CoE):** A CoE serves as a centralized hub for developing and scaling AI initiatives. It ensures alignment with corporate goals, provides expertise, and acts as a resource for other departments.

As an example, a large consumer electronics manufacturing company faced challenges with inconsistent AI applications across its global operations, resulting in inefficiencies in production and supply chain management. To address this, the company established an AI CoE to standardize AI practices, optimize production processes, and improve decision-making. The CoE developed governance frameworks, conducted regular performance audits, and ensured alignment with operational goals across all facilities. As a result, the company experienced a 30% increase in production efficiency through optimized predictive maintenance and quality control, a 25% reduction in downtime by proactively addressing equipment failures, and a 20% cost savings from streamlined operations and reduced resource waste. The CoE ultimately enabled more consistent, efficient AI applications, enhancing production capabilities and positioning the company for future growth.

- **Leveraging predictive analytics and real-time decision-making:** Large enterprises can use AI for predictive analytics and real-time decision-making, transforming operations across supply chain management, marketing, and customer service.

Expanded Example   A multinational clothing manufacturer implemented AI across its supply chain to reduce inefficiencies and improve forecasting accuracy. The company faced frequent supply chain disruptions, inaccurate demand forecasts, and excessive inventory levels, which increased costs and impacted production schedules.

The manufacturing company encountered several operational challenges:

- 25% inventory overstock, tying up capital and storage resources
- Frequent delivery delays, impacting production timelines and customer satisfaction
- Limited visibility into supplier performance and external factors affecting the supply chain

Solution:

The company implemented an AI-powered supply chain optimization platform by partnering with Blue Yonder. The platform integrated with the company's existing ERP and IoT systems to

- Analyze historical demand patterns, supplier data, and external factors like weather and geopolitical risks
- Generate real-time forecasts for raw material requirements
- Provide proactive alerts for potential supply chain disruptions

AI in action:

The AI system quickly identified inefficiencies and suggested optimizations:

- Highlighted excessive safety stock levels at regional warehouses
- Predicted potential shipment delays due to a supplier's production slowdown
- Recommended alternative suppliers, and optimized transportation routes to maintain production schedules

Outcome:

- 25% reduction in overstock and material waste
- 30% improvement in delivery timelines, enhancing production efficiency
- Enhanced supplier collaboration through shared predictive insights
- Improved customer satisfaction due to more reliable delivery schedules

The success of this initiative paved the way for broader AI adoption across other business units, including marketing and customer service.

Key Takeaways for Leadership

Tailoring AI strategy to the size of the organization is essential for success. For small businesses, this means focusing on affordable, scalable solutions that address immediate needs. Medium-sized companies must balance innovation with cost-efficiency and scalability. Large organizations, on the other hand, face the challenge of integrating AI into complex operations while ensuring ethical governance and cross-departmental coordination.

Not all AI investments provide immediate ROI. Some solutions deliver quick efficiency gains, whereas others require long-term strategic planning. The matrix presented in Table 4-2 categorizes AI investments into high-priority (immediate impact) versus long-term investments, based on company size.

Table 4-2: AI Investment Priorities Based on Company Size

BUSINESS SIZE	HIGH-PRIORITY AI INVESTMENTS	LONG-TERM AI INVESTMENTS
Small businesses	Chatbots, AI-powered CRMs, automation tools	Predictive analytics, supply chain AI
Medium businesses	AI-driven marketing, sales forecasting, logistics optimization	AI for cross-departmental collaboration, advanced AI analytics
Large enterprises	AI CoEs, predictive maintenance, real-time fraud detection	Fully autonomous AI-driven decision-making models

This matrix provides a structured way to evaluate AI investment strategies. Small businesses should prioritize AI-driven automation, and large enterprises must ensure that AI investments align with regulatory requirements and company-wide AI initiatives. Leaders should use this framework to balance short-term gains with long-term AI scalability.

A structured approach to AI implementation is critical for success. The roadmap presented in Figure 4-1 outlines the key steps businesses should take to deploy AI effectively, ensuring smooth integration and scalability as the company grows.

This roadmap provides a **structured blueprint** for AI adoption. Instead of rushing AI implementation, businesses should take an iterative approach—starting small, validating results, and scaling AI solutions gradually. Low-risk AI solutions include chatbots for small businesses,

AI sales forecasting for medium companies, and predictive analytics for large companies. With a well-defined roadmap, organizations can maximize AI's business impact while minimizing risks.

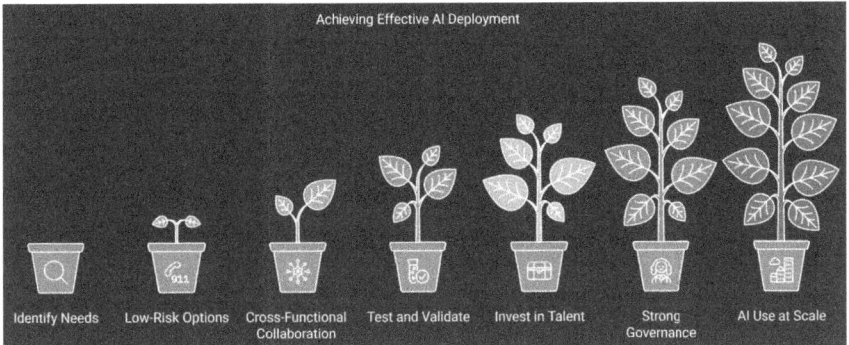

Figure 4-1: AI implementation roadmap based on company size

Regardless of company size, leadership must

- Align AI projects with strategic business goals.
- Encourage cross-functional collaboration.
- Invest in the right talent and partnerships.
- Establish strong governance to mitigate risks and ensure ethical AI deployment.

These topics and others will be covered in more depth in later chapters.

AI and Organizational Culture: Building an AI-First Mindset

Deploying Gen AI successfully requires more than technical expertise or well-crafted strategies—it demands a cultural transformation. Organizations must foster an environment where employees at all levels embrace AI as a tool for innovation, not as a threat to their roles. Building an **AI-first mindset** is about creating a workplace culture that values experimentation, collaboration, and trust in AI-driven decision-making.

This chapter explores how leaders can shape their organizational culture to ensure that Gen AI initiatives thrive. It offers actionable insights on overcoming resistance, promoting cross-functional collaboration, and embedding AI literacy into the fabric of the organization.

Why Culture Matters in AI Adoption

Although technology is a critical enabler of AI transformation, organizational culture ultimately determines its success or failure. Even the most advanced AI tools will fail to deliver a meaningful impact if employees resist their implementation, misunderstand their value, or feel disconnected from the change process. AI adoption is not just a technical shift—it's a cultural transformation that requires engagement, trust, and adaptability across all levels of the organization.

- **The role of culture in scaling AI:** A culture that supports AI adoption ensures smooth integration across departments, faster experimentation, and greater alignment with organizational goals. Conversely, a resistant culture can derail even the most promising projects.

- **Bridging strategy and execution:** Culture acts as the glue between an organization's AI strategy and its operational execution. It empowers employees to embrace change and take ownership of AI initiatives.

AI adoption is not just about technology—it's also about people. Many organizations struggle with resistance to AI, lack of trust, or siloed teams that prevent collaboration. Table 5-1 outlines the most common challenges businesses face when building an AI-first culture and provides actionable strategies to overcome them.

Table 5-1: AI Adoption Challenges and Solutions

CHALLENGE	DESCRIPTION	SOLUTION
Resistance to change	Employees fear job loss or disruption from AI.	Transparent communication, AI success stories, employee engagement.
Siloed teams	AI initiatives require cross-functional collaboration.	Create AI task forces, encourage knowledge sharing, reward collaboration.
Lack of trust in AI	Employees doubt AI's fairness or reliability.	Use explainable AI (XAI), involve employees in testing, ensure ethical AI policies.
Limited AI literacy	Employees lack basic AI knowledge or skills.	Provide AI training, role-specific learning paths, integrate AI education into onboarding.

Table 5-1 highlights the most common barriers to AI adoption and offers practical strategies to address them. Leaders who focus on transparent communication, cross-functional collaboration, and AI literacy will create an environment where AI is embraced, rather than feared.

Strategies for Building an AI-First Mindset

A number of strategies can be used to build an AI-first mindset. Let's look at four different strategies:

1. Overcoming resistance through transparent communication
2. Fostering cross-functional collaboration
3. Building trust in AI systems
4. Embedding AI literacy across the organization

1. Overcoming Resistance Through Transparent Communication

Resistance to AI often stems from fear and misunderstanding, making clear and proactive communication essential for fostering trust and enthusiasm. Employees may worry that AI will eliminate jobs, change roles, or introduce unfamiliar ways of working. Without proper guidance, these concerns can lead to skepticism and pushback, slowing adoption efforts.

To ensure a smooth transition, leaders must engage employees early, provide transparency, and demonstrate AI's tangible benefits. When employees understand why AI is being implemented, how it supports the company's mission, and how it benefits them personally, they are far more likely to embrace the change rather than resist it. The following strategies can help leaders create a culture of acceptance and excitement around AI initiatives:

- **Explain the "why":** Leadership must articulate how AI supports the company's mission and benefits employees. For example, if AI is being introduced to automate data entry, explain how this change will free employees to focus on more strategic and rewarding tasks. Use specific examples to show how similar changes have led to success in other organizations.

- **Highlight success stories:** Share real-world examples of how AI has positively impacted similar organizations. Use internal case studies if available. For instance, demonstrate how a pilot project within your own company streamlined a workflow or improved decision-making. These stories should be shared in meetings, newsletters, or company-wide town halls to build momentum.

- **Address fears proactively:** Employees may worry that AI will eliminate their jobs or make their work irrelevant. Create open forums, such as Q&A sessions or feedback surveys, where employees can voice concerns. Address these issues honestly, emphasizing that AI is a tool to augment their work rather than replace it.

Consider an example where a manufacturing company introduced predictive maintenance AI to reduce equipment downtime. Initially, workers feared job cuts, but leadership reassured them that AI would complement their expertise by reducing tedious troubleshooting tasks. Through transparent communication and training, employees embraced AI, resulting in a 30% efficiency improvement.

2. Fostering Cross-Functional Collaboration

AI initiatives are rarely confined to a single department. Successful AI adoption requires collaboration across IT, operations, marketing, customer service, and other business functions. Yet many organizations struggle with departmental silos, where teams operate independently, limiting data sharing and communication. These barriers can slow AI implementation, reduce effectiveness, and create fragmented AI strategies that fail to deliver full value.

To maximize AI's impact, leaders must foster a culture of collaboration, ensuring that insights, expertise, and data flow seamlessly across the organization. By breaking down silos and encouraging cross-functional teamwork, businesses can create AI strategies that are holistic, well-informed, and aligned with broader company goals. The following approaches can help organizations create a collaborative AI culture that accelerates innovation and adoption:

- **Create AI task forces:** Assemble teams with representatives from key departments, such as IT, marketing, operations, and customer service. These task forces should oversee pilot projects, ensuring that each department's needs and expertise are considered. Rotating leadership roles within the task force can also help create a sense of shared ownership.

- **Encourage knowledge sharing:** Use tools like Slack channels, collaborative software (e.g., Microsoft Teams, Asana), and internal wikis to facilitate knowledge exchange. Hold regular cross-departmental meetings to discuss AI learnings and insights, and invite team members to present successful initiatives.

- **Reward collaborative efforts:** Create incentive programs to reward teams that successfully collaborate on AI initiatives. Recognition can take the form of monetary bonuses, company awards, or public acknowledgment during meetings.

Let's look at an example of fostering cross-functional collaboration. A retail chain used an AI-powered recommendation engine to improve sales. By involving marketing, sales, and IT teams in the project, the company ensured seamless integration and alignment with customer engagement strategies. The initiative boosted conversion rates by 20%.

3. Building Trust in AI Systems

Trust is a cornerstone of successful AI adoption. Employees need to feel confident that AI tools are fair, reliable, and beneficial. To build trust in your AI systems, you can do the following:

- **Emphasize transparency:** Use XAI systems that provide clear reasoning behind decisions. For instance, if an AI-driven system recommends promotions, it should outline the criteria used, such as performance metrics or skill assessments. Regularly review AI outputs to ensure fairness and consistency.

- **Involve employees in testing:** Invite employees to test AI tools during the development phase. Their feedback can help refine the system, making it more user-friendly and aligned with their needs. This involvement also gives employees a sense of ownership over the tool's success.

- **Address bias and ethics:** Establish policies to prevent bias in AI systems, ensuring that data inputs are representative and decision-making processes are equitable. Offer training sessions on ethical AI use to help employees understand the company's commitment to fairness and transparency.

Trust is a key driver of AI adoption. Employees need to believe that AI is fair, transparent, and beneficial before they will fully embrace it. The framework shown in Figure 5-1 illustrates the core elements required to build trust in AI systems within an organization.

This framework provides a structured approach to AI trust-building. Organizations that implement explainable, fair, and reliable AI systems will see higher employee confidence, leading to greater AI adoption and effectiveness.

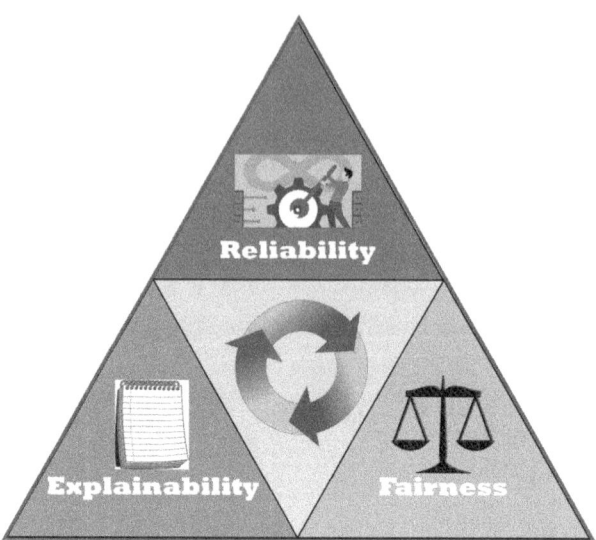

Figure 5-1: AI trust and transparency framework

For example, a financial institution implemented an AI-driven credit scoring system. To build trust, the company shared how the algorithm worked, the data it used, and steps taken to minimize bias. This transparency increased adoption among employees and customers alike.

4. Embedding AI Literacy Across the Organization

To embrace AI, employees need to understand its capabilities, limitations, and potential impact on their work. You can help your employees with this by embedding literacy across the organization and by doing the following:

- **Offer AI training for all levels:** Develop tiered AI training programs. For general staff, offer foundational courses covering AI basics, such as what AI is and how it applies to the organization. For technical staff, offer advanced workshops on deploying and managing AI tools.

- **Incorporate AI into onboarding:** Introduce new hires to the organization's AI initiatives during onboarding. This ensures that they understand AI's role in the company and their own responsibilities in leveraging it. Onboarding materials could include tutorials, case studies, and practical exercises.

- **Use role-specific learning paths:** Tailor training programs to the needs of specific teams. For example:

  - **Sales teams:** Teach them how to use AI-driven tools for lead scoring and customer segmentation.

  - **Operations teams:** Train them in predictive maintenance and workflow optimization.

  - **HR teams:** Show them how to use AI for recruitment and employee engagement.

AI literacy is not a "one-size-fits-all" approach. Different teams within an organization require different levels of AI training, depending on their role and responsibilities. Table 5-2 provides a breakdown of AI education needs based on job function.

Table 5-2: Example of AI Training and Literacy for Different Employee Roles

EMPLOYEE ROLE	AI TRAINING FOCUS	EXAMPLE TRAINING TOPICS
Executives and Leaders	AI strategy, ethical AI, business impact	AI-driven decision-making, governance, risk mitigation
Marketing and Sales	AI for customer engagement and personalization	AI-driven customer segmentation, predictive analytics
Operations and Supply Chain	AI for process automation and forecasting	Predictive maintenance, AI-powered logistics
HR and Talent Management	AI for hiring and workforce planning	AI-driven recruitment, performance prediction
IT and Data Teams	AI model development and deployment	Machine learning, AI security, data integration

Providing role-specific AI training ensures that employees understand how AI applies to their work. This targeted learning approach will increase AI fluency across all departments, allowing the company to maximize AI's value.

To illustrate embedding AI literacy, consider what one company did. A medium-sized tech firm introduced AI training modules for employees across departments. Sales teams learned how to use AI tools for customer segmentation, and HR used AI to streamline recruitment. This initiative boosted AI adoption rates by 40%.

Leadership's Role in Shaping Culture

Leadership sets the tone for an AI-first culture. To succeed, leaders must

- **Lead by example:** Leaders should actively use AI tools in their own workflows and showcase how these tools improve productivity. For example, a CEO might use an AI-driven dashboard during meetings to present real-time analytics, demonstrating its value to the organization.

- **Celebrate wins:** Recognize and reward teams and individuals who successfully implement AI initiatives. Highlight their work in newsletters, town halls, or team meetings to inspire others and reinforce the organization's commitment to AI.

- **Encourage experimentation:** Create a culture where employees feel safe experimenting with AI tools without fear of failure. Offer "innovation days" or hackathons where teams can brainstorm and prototype AI-driven solutions.

An AI-first culture doesn't happen overnight—it requires structured leadership initiatives that encourage AI adoption at all levels. The flowchart in Figure 5-2 outlines the key steps leaders can take to successfully integrate AI into the company culture.

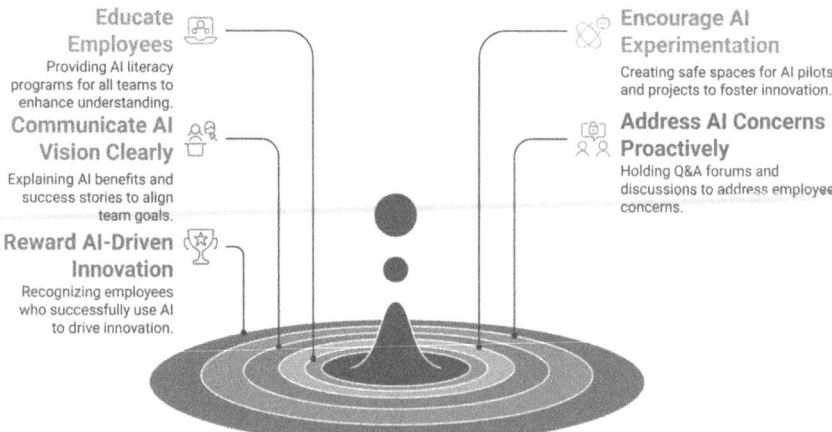

Figure 5-2: AI adoption roadmap for culture transformation

A well-defined roadmap helps organizations transition to an AI-first mindset. By following these steps, leaders can ensure that employees feel engaged, empowered, and ready to work alongside AI. For example, a CEO of a logistics company personally demonstrated the use of AI-driven dashboards during quarterly reviews, showcasing how real-time data improved decision-making. This visible endorsement motivated other leaders to adopt similar tools.

Laying the Foundation for AI Success

Building an AI-first culture is not a one-time initiative—it's an ongoing effort that requires commitment from leadership and engagement from employees. By fostering transparency, collaboration, trust, and literacy, organizations can create an environment where AI thrives as an enabler of innovation and efficiency.

As you progress in your AI journey, remember that culture is the foundation on which all AI strategies are built. A strong, AI-ready culture ensures that the tools and strategies discussed in the following chapters will not just succeed but create lasting value for your organization.

Building an AI-Ready Workforce

In today's rapidly evolving digital landscape, artificial intelligence (AI) is no longer a futuristic concept—it is a foundational driver of modern business transformation. As organizations shift toward an AI-first mindset, where AI is embedded in strategic decision-making and daily operations, the next critical step is ensuring that employees are equipped to work alongside AI effectively.

An AI-first strategy is not just about technology—it is about people. Whereas previous discussions have focused on why organizations must prioritize AI integration, this chapter explores how leaders can build an AI-ready workforce that can fully leverage AI's potential. Successfully deploying AI requires more than implementing new tools; it demands visionary leadership, a commitment to continuous learning, and a culture that embraces change. Organizations that fail to prepare their employees for AI risk slow adoption, resistance, and inefficiencies, ultimately limiting the technology's impact.

Preparing employees for AI involves upskilling and reskilling initiatives, fostering collaboration across departments, and creating a work environment where AI is seen as an enabler rather than a threat. Leaders play a crucial role in this transformation by setting the tone for AI adoption, ensuring employees feel supported, and creating opportunities for them to learn, adapt, and thrive in an AI-enhanced workplace. This chapter will explore the key strategies for developing an AI-ready workforce, including skills development, change management, and cross-functional collaboration, all of which are essential for organizations aiming to scale AI successfully and sustain long-term competitive advantage.

Leadership's Role in Workforce Preparation

A key responsibility of leadership is ensuring that employees are ready to meet the demands of an AI-driven future. This involves not only identifying the skills gaps in the workforce but also leading the charge in developing the competencies required to thrive in an AI-enabled workplace. This includes doing things such as

- **Identifying critical AI skills:** Leadership should begin by identifying the critical skills that will be needed as AI becomes integrated into business operations. These may include technical competencies, such as data analysis, machine learning, and programming, but also essential soft skills such as problem-solving, critical thinking, and adaptability. Leadership must take a hands-on approach to conducting a workforce skills assessment to identify gaps and opportunities for growth.

- **Developing AI literacy for all employees:** AI literacy is not just for technical staff—it is a key component of an AI-ready workforce at all levels of the organization. Leaders should spearhead initiatives that introduce employees to basic AI concepts and practical applications relevant to their roles. This foundational knowledge enables nontechnical employees to understand how AI impacts their work and fosters a collaborative environment where employees across departments can work with AI effectively.

Preparing employees for AI adoption requires a mix of technical and soft skills. Table 6-1 outlines the key competencies needed for an AI-ready workforce and the role of leadership in fostering these skills.

Table 6-1: AI Skills and Competencies for an AI-Ready Workforce

SKILL CATEGORY	SKILL	WHY IT'S IMPORTANT	EXAMPLE USE CASE
Technical skills	Data literacy	Understanding, analyzing, and interpreting data	Using AI dashboards for business insights
	Machine learning basics	Knowing AI's capabilities and limitations	Collaborating with AI developers
Soft skills	Critical thinking	Evaluating AI-generated recommendations	AI-assisted decision-making
	Adaptability	Embracing AI-driven changes	Transitioning to AI-enhanced roles
Ethical and governance skills	AI ethics awareness	Understanding bias, transparency, and fairness	Identifying unethical AI practices
	AI governance implementation	Ensuring responsible AI usage	Creating AI compliance policies

AI literacy is **not just for technical teams**—it is essential for employees across all departments. By fostering these skills, leadership can ensure that AI becomes an enabler of success rather than a source of resistance.

As an example, a CEO might introduce AI basics workshops across all departments, focusing on how AI can be used in customer service, marketing, or supply chain operations. This gives employees a clear understanding of how AI will enhance their daily tasks, reducing fear and resistance to AI adoption.

Ethics and Governance as Core Leadership Responsibilities

As organizations prepare to adopt AI at scale, leadership must go beyond developing technical expertise and focus on establishing a strong ethical and governance framework. AI technologies have immense power to drive innovation, but they also present risks—such as biases in decision-making, privacy concerns, and ethical accountability. Without a structured approach to ethics and governance, AI adoption can lead to unintended consequences that erode trust, create legal vulnerabilities, and hinder long-term success.

To navigate these challenges, leaders must define and enforce ethical principles that guide AI development and use. This means embedding ethics into AI training programs, ensuring that employees understand both the opportunities and risks associated with AI, and creating policies that promote fairness, transparency, and accountability.

Defining Ethics, Governance, and Structure in AI Leadership

Before diving deeper, it is essential to define the core concepts that underpin responsible AI leadership:

- **Ethics** in AI refers to the principles and moral considerations that guide AI development and deployment. Ethical AI ensures that AI systems operate fairly, transparently, and without harm to individuals or society. Ethics in AI includes topics such as bias mitigation, data privacy, fairness, and the accountability of AI-driven decisions.

- **Governance** refers to the frameworks, policies, and mechanisms that regulate AI systems within an organization. AI governance ensures that AI operates within legal, regulatory, and ethical boundaries and provides oversight mechanisms to prevent misuse. Governance is what transforms ethical principles into enforceable guidelines that shape AI decision-making.

- **Structure** in the context of AI governance refers to the formalized systems, policies, and oversight mechanisms that ensure ethical AI implementation. A well-defined structure provides clear roles and responsibilities for AI oversight, ensuring that employees and leadership understand who is accountable for ethical AI decisions and how policies are enforced.

Together, ethics define the principles, governance establishes the rules, and structure provides the mechanism for enforcing them. Strong leadership in AI requires all three elements to be integrated into the organization's AI adoption strategy.

The Importance of AI Governance

AI governance provides the necessary structure for developing and using AI systems responsibly. It involves setting clear guidelines for how AI models are trained, evaluated, and deployed to ensure that they align with both the organization's ethical standards and external regulatory requirements.

Leaders must take an active role in defining these governance frameworks, ensuring that AI policies are enforceable, transparent, and adaptable to evolving technologies. This means implementing policies for responsible data usage, bias detection, transparency, and accountability. Strong AI governance protects organizations from reputational, legal, and ethical risks while also fostering trust and reliability in AI-driven decision-making.

For example, a well-structured governance framework should include

- **Data ethics:** AI models rely on vast amounts of data, making ethical data collection and usage paramount. Governance structures should ensure that all data used for AI training complies with privacy laws and respects user consent. Employees must be trained to understand the legal and ethical boundaries of using personal data, particularly in high-risk AI applications like healthcare, finance, and hiring.

- **Bias mitigation:** AI models can inadvertently reinforce biases present in training data, leading to discriminatory outcomes. Governance frameworks should include mandatory audits, diverse dataset requirements, and ongoing bias testing to identify and mitigate potential biases in AI models. Leaders must also ensure that employees are educated on bias risks and understand their role in developing fair AI systems.

- **Transparency and explainability:** AI-driven decisions should be understandable, auditable, and explainable. This is especially critical in regulated industries like finance, healthcare, and employment, where AI outputs directly impact people's lives. Governance structures should require AI models to generate human-readable explanations for their decisions, and organizations should cultivate a culture where AI transparency is treated as a core responsibility, not an afterthought.

Upskilling for Ethical AI

As AI continues to transform the workplace, upskilling and reskilling employees is essential for organizations to stay competitive. Leaders must take an active role in fostering a culture of continuous learning, ensuring that employees are not only encouraged but also incentivized to expand their knowledge and develop new skills. Leadership-driven upskilling initiatives should include comprehensive training programs that address both technical AI competencies and essential soft skills, enabling employees to navigate a rapidly changing work environment. Upskilling efforts should prepare employees to take on AI-enhanced roles, and reskilling programs should focus on helping those whose roles may be automated transition to new opportunities within the organization.

Beyond structured training programs, fostering a learning culture is equally critical. Leaders must set the tone for lifelong education, reinforcing the idea that professional growth is not optional but a fundamental part of adapting to AI-driven transformation. Organizations can cultivate this mindset by providing financial support for courses, allowing time off for training, and offering in-house development programs that enable employees to expand their skill sets without disrupting their work–life balance. By prioritizing education and creating an environment where employees feel empowered rather than threatened by AI, leaders can ensure that their workforce remains adaptable, innovative, and fully equipped for the future.

Leadership's Role in Ethical AI

Ethical leadership is pivotal in ensuring that the principles of governance and ethical AI are woven into the fabric of organizational culture. Leaders must set the tone by championing ethical practices and making it clear that cutting corners on governance or ethics is unacceptable, even in pursuit of rapid innovation.

Key leadership responsibilities include

- **Creating ethical AI guidelines:** Leaders must work with data scientists, legal teams, and ethicists to develop comprehensive guidelines that govern the use of AI across the organization.

- **Fostering an open dialogue on ethics:** Regular discussions on AI ethics should be encouraged within teams. Leaders should create platforms for employees to raise concerns or seek clarification on the ethical implications of their AI projects.

- **Accountability:** Leaders must ensure that employees understand the importance of accountability in AI systems. Establishing clear reporting channels, regular audits, and assigning specific roles for ethical oversight can help reinforce this responsibility throughout the workforce.

Change Management and Employee Engagement

AI adoption often leads to significant changes in processes, workflows, and roles. For AI implementation to succeed, leadership must take charge of managing these changes effectively. Employees may feel uncertain about how AI will impact their roles, and it is the responsibility of leaders to guide them through the transition.

AI adoption often sparks resistance and uncertainty among employees. Leaders must manage this transition carefully through clear communication and engagement strategies. The flowchart in Figure 6-1 outlines leadership's role in AI change management.

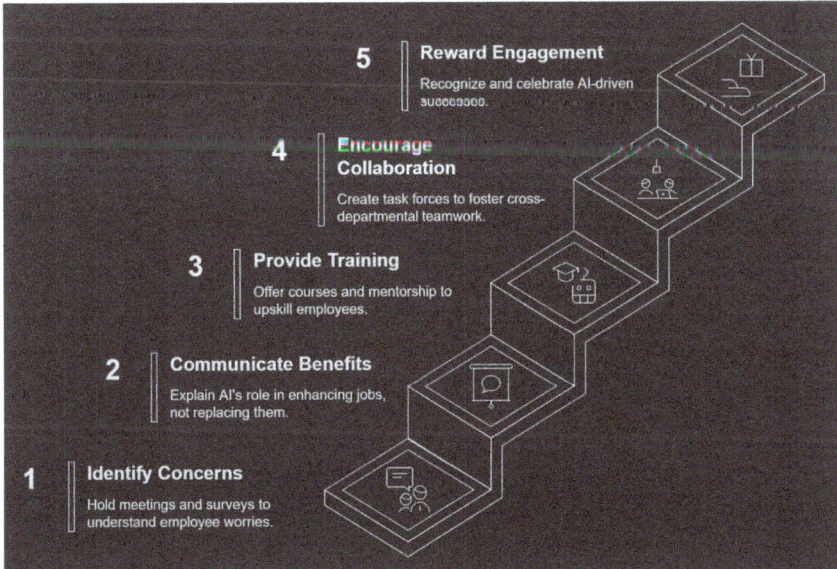

Figure 6-1: Leadership's role in AI workforce change management

Leadership must manage resistance to AI. It is natural for employees to feel anxious about the impact of AI on their jobs, particularly if they fear displacement. Leadership should address these concerns head-on by fostering open communication and transparency. Employees need to understand that AI is meant to augment, not replace, their roles and that their growth within the organization is a priority.

Additionally, leadership must build trust, which they can do through communication. Transparent leadership is crucial when building an AI-ready workforce. Leaders should communicate AI's strategic importance clearly and consistently, outlining how it will drive business value and improve employee outcomes. By framing AI as a tool that empowers employees to focus on higher-value tasks, leaders can build trust and encourage employees to embrace AI rather than resist it.

Successful AI transformation depends on employee trust and engagement. By following the change management framework presented in Figure 6-1, leaders can build a workforce that is not just AI-ready, but AI-empowered.

Consider what the leadership at a large manufacturing firm did. They introduced a series of "town hall" style meetings to discuss AI integration plans. Employees were encouraged to ask questions, voice concerns, and provide input on how AI might affect their day-to-day work. This open dialogue reduced anxiety and resistance, as employees felt heard and involved in the AI transformation process.

Promoting Collaboration Between Departments

AI integration often requires cross-functional collaboration between technical and nontechnical teams. Leadership plays a critical role in breaking down silos and fostering a culture of collaboration, where employees from different departments work together to implement AI-driven solutions.

Leadership needs to be involved in the development of cross-functional teams. Leaders should encourage the formation of cross-functional AI implementation teams that bring together data scientists, engineers, business analysts, and operational staff. This ensures that AI solutions are developed and applied in ways that benefit the entire organization, rather than just isolated departments.

For example, in a global consumer goods company, leadership established cross-functional teams made up of employees from IT, supply chain management, and sales to work on AI-driven demand forecasting. This collaborative approach ensured that AI systems were designed to meet the needs of both the technical and business sides of the company, leading to a 15% improvement in forecasting accuracy.

Creating a Roadmap for AI Workforce Transformation

Leadership must be strategic and forward-thinking when developing a roadmap for long-term workforce transformation. Unlike fostering an AI-first culture, which focuses on integrating AI into decision-making and organizational strategy, workforce transformation is about ensuring that employees have the skills, knowledge, and adaptability to work alongside AI effectively. Whereas an AI-first culture prioritizes AI-driven thinking across leadership and operational frameworks, workforce development is a practical, structured process that equips employees with the necessary tools to succeed in an AI-enhanced environment.

Visionary Leadership in Workforce Transformation

Leaders must establish a clear vision for how the workforce will evolve alongside AI, ensuring that employees see AI as an opportunity rather than a disruption. This requires structured career development programs, clearly defined pathways for transitioning into AI-driven roles, and ongoing support for learning and skill development. A successful roadmap does not simply train employees on existing AI applications but prepares them for future advancements in AI technology, ensuring long-term workforce resilience and adaptability.

A Roadmap for AI Workforce Development

An AI-ready workforce is built over time through structured leadership initiatives that align workforce transformation with organizational AI adoption. The roadmap outlined in Figure 6-2 provides a step-by-step approach for integrating AI into workforce development, focusing on progressive skill-building, role transitions, and leadership-driven education programs.

A well-structured roadmap ensures that AI adoption is gradual, inclusive, and sustainable. Unlike an AI-first culture, which establishes AI as a core component of business strategy, workforce transformation ensures that employees remain engaged, empowered, and fully prepared to work within an AI-driven ecosystem. Companies that take a proactive and structured approach will not only mitigate workforce displacement concerns but also create a highly skilled, AI-ready team that drives long-term business success.

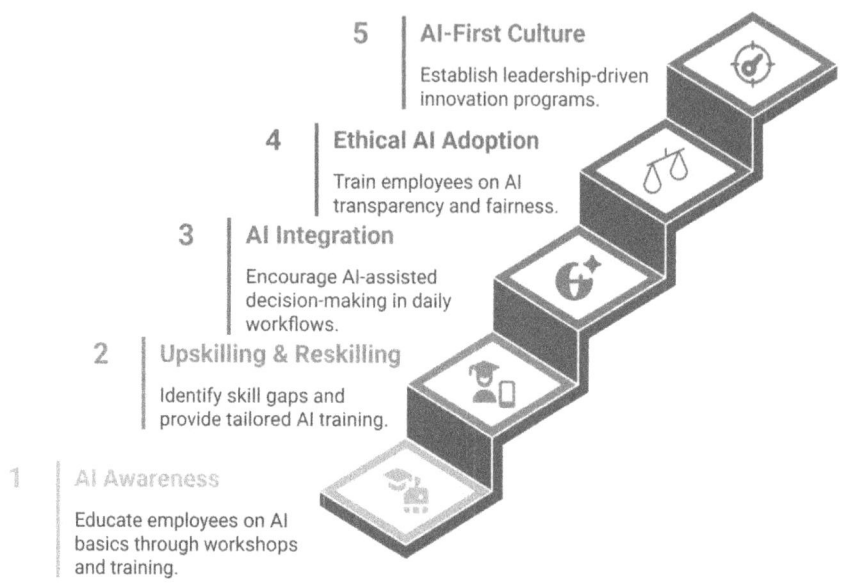

5 AI-First Culture
Establish leadership-driven innovation programs.

4 Ethical AI Adoption
Train employees on AI transparency and fairness.

3 AI Integration
Encourage AI-assisted decision-making in daily workflows.

2 Upskilling & Reskilling
Identify skill gaps and provide tailored AI training.

1 AI Awareness
Educate employees on AI basics through workshops and training.

Figure 6-2: AI workforce transformation roadmap

Consider a manufacturing company implementing AI-driven automation in its production processes. Instead of simply adopting AI tools and expecting employees to adjust, leadership developed a structured roadmap for workforce transformation. They introduced an internal training program that provided workers with certifications in AI systems management and robotics, allowing employees to transition from manual labor roles to more advanced technical positions. Over time, this approach ensured that the workforce evolved alongside AI, rather than being displaced by it, enabling the company to maximize AI's potential while retaining experienced employees.

By differentiating workforce transformation from AI-first culture, leaders can align AI strategy with workforce readiness, ensuring that AI adoption is not only a technological shift but a human-centered evolution that benefits the entire organization.

Case Study: A Manufacturing Company Prepares Its Workforce for AI

A large manufacturing company in the automotive industry recognized the potential of AI to improve its operations, particularly in supply chain management and production efficiency. However, the company also knew that its workforce would need significant upskilling to successfully integrate AI into its processes.

Leadership launched a comprehensive upskilling initiative focused on AI literacy, robotics, and data analysis. The company introduced training programs in partnership with local technical schools and offered employees time off to complete certification courses. Leadership also communicated openly about the long-term strategy for AI adoption, emphasizing how AI would enhance roles rather than replace them.

As a result, employees across the organization felt more confident in their ability to work alongside AI, and the company saw a 20% increase in production efficiency. By taking a leadership-driven approach to workforce transformation, the company successfully prepared its employees for the future of AI while maintaining a collaborative and growth-oriented culture.

The Critical Role of Leadership in Building an AI-Ready Workforce

Building an AI-ready workforce is not just about training employees in technical skills; it's about leadership's ability to guide the organization through change, foster a culture of learning, and inspire employees to see AI as a tool for empowerment. By investing in upskilling, promoting collaboration, and communicating a clear vision for the future, leaders can ensure that their organizations are fully prepared to harness the power of AI.

Practical Steps for Implementing Gen AI Across the Organization

Successfully integrating Gen AI across an organization involves more than just adopting new technology. Like previous rollouts of technology, getting its true value requires strategic planning, targeted pilot projects, a solid infrastructure, and a clear pathway for scaling and continuous improvement. By following a structured approach, leaders can maximize AI's impact while minimizing risks. In the following sections, I present such a structured approach using five steps:

1. Assessing business needs
2. Conducting AI pilot projects
3. Building the right infrastructure and selecting tools
4. Scaling AI solutions across the organization
5. Continuous measurement and refinement

Assessing Business Needs

Before implementing AI, it is crucial for organizations to determine where the technology will have the greatest impact. A thoughtful assessment of business needs is essential for prioritizing projects, aligning AI initiatives with organizational goals, and identifying areas where AI can effectively address existing challenges.

Successful AI implementation hinges on clearly linking AI efforts to strategic business objectives, such as enhancing customer satisfaction, improving operational efficiency, or driving revenue growth. By ensuring that AI is connected to these overarching goals, companies can maximize its potential and deliver tangible results. Table 7-1 outlines key factors organizations should evaluate to ensure a successful AI deployment. The Evaluation column allows you do a quick check to see where your organization lies with respect to readiness for AI scalability.

Table 7-1: AI Readiness Assessment Framework

CATEGORY	ASSESSMENT CRITERIA	WHY IT MATTERS	EVALUATION (✔ / ✘)
Data readiness	Is the data structured and high-quality?	AI requires clean, well-organized data to function effectively.	
	Are privacy and security measures in place?	AI models must comply with data protection regulations.	
Technology and infrastructure	Is the IT infrastructure scalable for AI workloads?	AI requires significant processing power and storage.	
	Cloud vs. on-premises: which approach fits best?	Cloud solutions provide flexibility; on-premises offers data control.	
Workforce and talent	Do employees have the necessary AI skills?	AI adoption success depends on AI literacy and training.	
	Is leadership committed to AI initiatives?	Executive buy-in ensures strategic alignment.	

One of the first steps in the AI readiness assessment process is analyzing the pain points within the organization. These are the inefficiencies, high costs, and process bottlenecks that AI is well-suited to solve. For example, AI may be able to streamline operations, reduce manual workloads, or optimize decision-making in areas where traditional processes are proving inadequate. Identifying these pain points helps narrow the focus to areas where AI will have the most significant impact.

Additionally, the success of AI projects heavily depends on the quality and availability of data. Organizations need to evaluate the state of their data across departments, ensuring that the data is robust, structured, and ready to support AI-driven initiatives. Without a solid data foundation, even the most sophisticated AI models may fail to deliver the expected outcomes.

To kick off the AI implementation process, organizations should engage key stakeholders in cross-departmental workshops. These workshops allow department heads and other leaders to discuss areas where AI could drive improvements. By exploring how AI might address specific challenges within functions like marketing, operations, or customer service, organizations can build a comprehensive view of where AI would be most beneficial.

Once potential AI use cases are identified, the next step is to prioritize them based on factors such as feasibility, expected return on investment (ROI; covered in later chapters), and alignment with the overall business strategy. Using a structured scoring system can help rank these projects, allowing leaders to focus on those that promise the most immediate or significant results. This ensures that AI efforts are directed toward initiatives that not only solve pressing issues but also deliver quick wins to build momentum and support within the organization.

Let's look at a scoring criteria example. The following outlines an AI project scoring criteria and presents an example assessment in Table 7-2.

Table 7-2: Scoring Assessment Example

PROJECT	BUSI- NESS IMPACT	STRA- TEGIC ALIGN- MENT	FEASI- BILITY	RESOURCE AVAIL- ABILITY	TIME- TO- VALUE	INNO- VATION POTEN- TIAL	TOTAL SCORE
Marketing AI	5	5	4	3	5	4	26
Production AI	3	3	5	4	3	3	21
AI chatbot	4	4	3	5	2	2	20

Scoring Criteria

1. **Business impact (30%)**

   - **High (5 points):** Significant cost savings, revenue generation, or risk reduction
   - **Medium (3 points):** Moderate improvement to efficiency or productivity
   - **Low (1 point):** Minor incremental improvement

2. **Strategic alignment (20%)**

   - **High (5 points):** Directly aligns with company strategy and long-term vision
   - **Medium (3 points):** Indirect alignment but still supports key goals
   - **Low (1 point):** Little alignment with strategic priorities

3. **Feasibility and readiness (20%)**

   - **High (5 points):** Technically viable, minimal roadblocks, required data available
   - **Medium (3 points):** Some challenges but solvable with reasonable effort
   - **Low (1 point):** Significant feasibility concerns or major unknowns

4. **Resource availability (15%)**

   - **High (5 points):** Required personnel, tools, and funding are available
   - **Medium (3 points):** Some key resources are available but require adjustments
   - **Low (1 point):** Major resource constraints exist

5. **Time-to-value (10%)**

   - **Short-term (5 points):** Results expected within 3–6 months
   - **Mid-term (3 points):** Results expected in 6–12 months
   - **Long-term (1 point):** Takes over a year to deliver value

6. **Innovation potential (5%)**

   - **High (5 points):** Game-changing, industry-leading innovation
   - **Medium (3 points):** Incremental improvement but differentiates the business
   - **Low (1 point):** Standard industry practice

Finally, organizations must conduct an AI readiness assessment. This involves evaluating the current state of their data infrastructure, technological capabilities, and talent pool to determine whether they are prepared for AI adoption. This assessment helps identify gaps in the organization's tools, data resources, or workforce that need to be addressed before fully embarking on AI projects. By understanding where improvements are needed—whether in investing in new technology, sourcing additional data, or upskilling employees—organizations can ensure that they are well-equipped to implement AI effectively.

Conducting AI Pilot Projects

Pilot projects provide an effective way for organizations to test the potential of AI solutions on a small scale before committing to a full-scale rollout. By starting with manageable, controlled experiments, companies can evaluate AI's real-world impact, measure its effectiveness, and refine the technology based on insights gained during the trial period. This approach allows for informed decision-making about whether to scale AI across the organization, reducing the risks associated with widespread implementation.

The first step in any pilot project is to carefully define its scope. A well-chosen pilot should be narrow enough to allow for close monitoring and adjustments but impactful enough to demonstrate the true potential of AI within the organization. It is important to select a project that provides tangible benefits and can showcase AI's capabilities without overwhelming resources. This focused approach ensures that the organization can learn from the experience and fine-tune the technology for broader deployment.

Another critical factor in a successful AI pilot is cross functional collaboration. AI solutions often involve multiple parts of the organization, including IT, data science, and business units that will directly interact with the technology. Bringing these teams together early in the pilot process ensures that all perspectives are considered and potential roadblocks are addressed. This collaborative approach helps align the AI initiative with broader business goals and operational realities, increasing the likelihood of long-term success.

Equally important is setting clear metrics for success. Key performance indicators (KPIs) should be established from the outset to provide measurable benchmarks for the pilot's performance. For example, if the pilot is focused on AI-driven customer service, KPIs might include metrics such as reduced response times, improved customer satisfaction, or the percentage of inquiries successfully handled by the AI. These KPIs not only demonstrate the project's value but also provide data-driven evidence to support further investment in AI.

To maximize impact, organizations should ensure that the pilot aligns with strategic business goals and has well-defined objectives. For instance, if the goal is to improve customer interactions, a chatbot or AI-based recommendation engine could be piloted to test how well it meets those objectives. Establishing success metrics tied directly to the pilot's goals ensures that results can be effectively measured and used to drive decision-making. Table 7-3 provides sample KPIs that can help assess AI performance in different business functions.

Table 7-3: Key Performance Indicators for AI Pilots

AI APPLICATION	KPIS	WHY IT'S IMPORTANT
AI-powered customer service chatbot	Response time reduction, % of inquiries resolved without human intervention, customer satisfaction (CSAT) increase	Measures AI's impact on efficiency and customer experience
AI-driven supply chain optimization	Reduction in inventory costs, % decrease in delays, demand forecasting accuracy	Demonstrates how AI improves logistics and forecasting
AI-assisted fraud detection	% of fraudulent transactions identified, false positive rate, investigation time reduction	Evaluates AI's accuracy and ability to reduce financial losses

Setting clear KPIs ensures that AI pilots deliver measurable business impact, helping organizations decide whether to scale AI solutions.

Throughout the pilot, it's essential to monitor progress and collect feedback from all stakeholders involved—both employees who work with the AI and customers who interact with it. Continuous monitoring against the established KPIs ensures that any issues are identified and addressed early. Feedback from employees and users provides valuable insights into areas of strength and weakness, allowing for adjustments to the AI solution before it is scaled. By refining the system based on real-world use, organizations can ensure that when the AI is rolled out on a larger scale, it delivers optimized results.

In summary, pilot projects are an invaluable tool for testing AI in a controlled environment, helping organizations gather crucial insights, measure impact, and make data-driven decisions about broader deployment. With a focused scope, cross-departmental collaboration, and clear success metrics, businesses can confidently scale AI solutions that have been proven to deliver value.

Building the Right Infrastructure and Selecting Tools

For organizations to successfully deploy AI solutions, it is essential to build the right infrastructure that supports data storage, processing, and AI tools. A robust infrastructure not only ensures that AI models run efficiently but also allows them to scale seamlessly as the organization grows. The choice between cloud-based and on-premises AI infrastructure has far-reaching implications. Cloud solutions such as AWS, Google Cloud, and Microsoft Azure provide elastic scalability, rapid deployment, and prebuilt AI tools that allow organizations to iterate quickly and expand their AI capabilities without needing to invest in extensive hardware. Cloud platforms also offer built-in security features, regulatory compliance tools, and AI model hosting services, making them an ideal choice for companies that prioritize flexibility and speed. The diagram in Figure 7-1 compares cloud-based and on-premises AI deployment to help organizations choose the best fit.

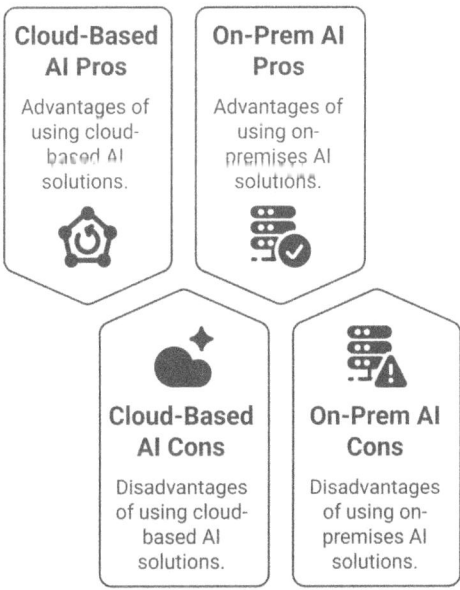

Figure 7-1: Cloud versus on-premises AI considerations

On-premises AI infrastructure may be more suitable for organizations with strict data privacy regulations, security concerns, or legacy system dependencies. On-premises environments give companies full control over data storage, processing, and access management, which is especially critical in industries such as finance, government, and healthcare, where regulatory compliance is a top priority. The drawback, however, is that on-premises AI infrastructure requires significant capital investment, ongoing maintenance, and dedicated IT expertise to manage hardware, software, and security updates.

For many organizations, a hybrid approach is the best option. A hybrid AI infrastructure allows companies to store and process sensitive data on-premises while leveraging the cloud's computational power and AI tools for model training, deployment, and analytics. This approach enables organizations to balance security with scalability, ensuring that AI models can run efficiently while adhering to industry regulations.

Selecting the Right AI Platforms and Tools

Once an organization's infrastructure is in place, the next step is selecting the right AI platforms and tools that align with business objectives. AI platforms must provide

- **Seamless integration with existing enterprise systems** (enterprise resource planning [ERP], customer relationship management [CRM], supply chain platforms)

- **Robust security and compliance features** to meet regulatory standards

- **Scalability** to accommodate future AI model expansion and increased data loads

- **Prebuilt AI models and development environments** that accelerate AI adoption

Beyond selecting a cloud provider, organizations should assess specialized AI platforms that cater to specific industry needs, such as IBM Watson for healthcare AI, OpenAI's API for conversational AI, and TensorFlow or PyTorch for deep learning applications.

Additionally, organizations must invest in data engineering tools that enable efficient data management, preprocessing, and cleaning, ensuring that AI models receive high-quality inputs. Without well-structured data pipelines, even the most powerful AI models will produce unreliable results.

Establishing Strong AI Governance and Security Protocols

Infrastructure and tools alone are not enough: data governance and security play a crucial role in AI success. AI governance refers to the policies, procedures, and ethical guidelines that dictate how AI systems handle data, make decisions, and remain transparent. Organizations must implement

- Data privacy protocols that comply with the General Data Protection Regulation (GDPR), California Consumer Privacy Act (CCPA), and industry-specific regulations
- Bias-mitigation frameworks to prevent AI from amplifying discriminatory patterns
- Audit trails and explainability mechanisms that allow AI decision-making processes to be monitored and reviewed

AI security is also paramount. Organizations must protect AI models from adversarial attacks, data breaches, and model drift, ensuring that AI-generated insights remain trustworthy. Regular security audits, encryption methods, and access controls should be integrated into AI governance frameworks to prevent unauthorized data manipulation and misuse.

By building a scalable, secure, and well-governed AI infrastructure, organizations lay the foundation for seamless AI deployment and sustainable long-term adoption. This will be covered more in depth in Chapter 10.

Scaling AI Solutions Across the Organization

Once AI pilot projects have demonstrated success, the next challenge is expanding AI adoption across multiple departments and business units. Scaling AI involves standardizing processes, integrating AI models into day-to-day workflows, and ensuring that employees across the organization can effectively work with AI systems. Table 7-4 highlights common barriers to AI adoption and proven solutions to address them.

Creating Standardized AI Workflows and Best Practices

AI adoption often starts as isolated experiments in specific departments, but for AI to create enterprise-wide impact, it must be embedded into business processes in a consistent and repeatable manner. Organizations should develop standard operating procedures (SOPs) for AI implementation, defining

- How AI models are deployed, updated, and maintained to ensure consistent performance
- Data input and validation protocols to maintain AI accuracy across different teams
- User training guides that explain how employees should interact with AI-powered systems

By developing AI Centers of Excellence (CoEs), organizations can create dedicated teams responsible for developing best practices, training staff, and monitoring AI adoption across departments. This ensures that AI models remain aligned with business goals and do not become isolated within a single function.

Managing Organizational Change and AI Adoption Challenges

One of the biggest barriers to scaling AI is resistance from employees who fear AI will replace their jobs or introduce complexity into their existing workflows. Change management is critical for overcoming these challenges.

Leaders must communicate AI's role as an enabler rather than a disruptor, emphasizing that AI is designed to enhance productivity, automate repetitive tasks, and free up employees for higher-value work. Organizations should

- Create AI training programs that help employees understand AI's capabilities and applications
- Establish feedback loops where employees can express concerns and provide insights into AI adoption
- Recognize and reward AI-driven initiatives to reinforce positive engagement with AI tools

By treating AI as a partnership between humans and technology, organizations can foster a culture of AI acceptance and collaboration, ensuring that AI solutions are fully embraced at every level. Table 7-4 highlighted common barriers to AI adoption and proven solutions to address them.

Table 7-4: Scaling AI: Common Challenges and Solutions

CHALLENGE	WHY IT HAPPENS	SOLUTION
Resistance to AI adoption	Employees fear job displacement and AI complexity.	Provide AI literacy training, focus on AI–human collaboration.
Siloed AI initiatives	AI adoption is fragmented across departments.	Create AI governance committees and standard operating procedures.
Lack of AI talent	Shortage of skilled AI professionals within the company.	Partner with AI vendors, invest in internal upskilling programs.

In summary, scaling AI across an organization is a strategic process that involves standardizing procedures, managing change, and providing continuous training and support. By creating a detailed scaling plan, establishing clear SOPs, and ensuring that employees are equipped to work with AI, organizations can successfully expand AI solutions to drive greater efficiency and innovation throughout the business.

Continuous Measurement and Refinement

Implementing AI is not a one-time effort but an ongoing process that requires continuous monitoring, feedback collection, and adjustments. For AI solutions to remain effective and relevant to evolving business needs, organizations must regularly assess their performance and make refinements as necessary. This process ensures that AI systems maintain high levels of accuracy, reliability, and alignment with organizational goals over time. Without this continuous attention, AI systems may degrade in quality or fail to keep pace with shifting business environments and data patterns.

One of the most critical aspects of maintaining AI effectiveness is real-time monitoring. AI solutions need to be constantly tracked to assess their performance across various metrics, such as accuracy, response times, and other KPIs specific to the application. Real-time monitoring allows organizations to detect issues as soon as they arise, enabling prompt interventions to prevent larger problems. Monitoring tools provide dashboards and alerts that allow teams to stay informed about the health and performance of AI systems at all times, ensuring that the AI continues to operate at optimal levels.

In addition to technical monitoring, organizations should also estab-lish feedback loops to collect input from users and stakeholders who interact with the AI system. This feedback is invaluable for identifying areas where the AI solution could be improved or refined to better meet business needs. By gathering insights from employees, customers, and other stakeholders, organizations can pinpoint weaknesses in the system, such as functionality that needs improvement or areas where the AI's outputs may not fully align with business goals. This iterative feedback process helps ensure that AI systems remain aligned with real-world demands and continue to deliver value.

Retraining and updating models is another essential aspect of main-taining AI effectiveness. As new data becomes available and business environments change, AI models may require retraining to ensure that they remain accurate and relevant. Regular updates based on shifting data patterns or business priorities are critical, particularly in dynamic environments where data changes frequently. Scheduling periodic model retraining sessions ensures that the AI system adapts to new information, preventing it from becoming obsolete or inaccurate over time.

AI success is not a one-time event—it requires continuous moni-toring, feedback collection, and refinement. The framework presented in Figure 7-2 outlines how organizations can ensure that AI models remain accurate, reliable, and aligned with business goals.

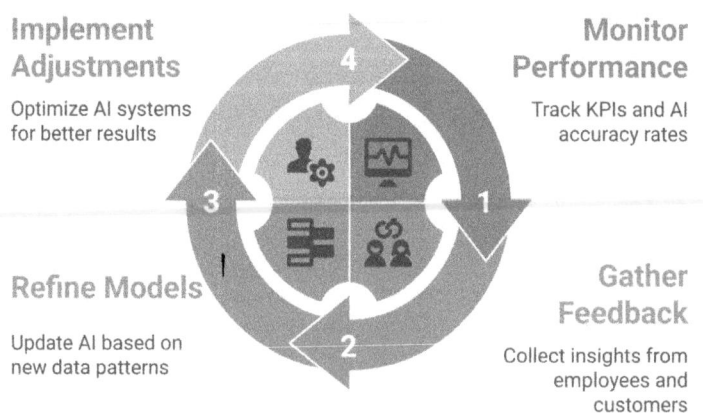

Figure 7-2: Continuous AI improvement framework

Organizations that commit to ongoing AI refinement ensure long-term value creation, business adaptability, and sustainable AI success. To begin with, organizations should implement robust monitoring systems to track the performance of their AI solutions in real time. These systems allow for continuous oversight of critical metrics and help teams respond quickly to any issues that arise. Dashboards and alert systems provide visibility into how the AI is functioning, offering detailed insights into performance metrics such as accuracy rates, response times, and error rates. This enables proactive management of AI systems, reducing the likelihood of performance degradation.

Next, gathering and analyzing feedback from users who interact with the AI system is crucial. Regularly collecting input from employees, customers, and other stakeholders helps identify any discrepancies between the AI's intended functionality and its actual performance in the real world. Feedback can reveal areas where the AI might not be delivering as expected, allowing the organization to make targeted improvements. By analyzing this feedback, organizations can refine the system to better meet the needs of its users and stakeholders.

Finally, retraining and refining AI models should be a regular practice. As new data comes in and business conditions evolve, AI models must be updated to reflect these changes. AI systems that are not periodically retrained may begin to lose accuracy, especially in environments where data is constantly shifting. Scheduling regular model retraining sessions ensures that AI systems remain accurate and reliable, continuously learning from new data to improve performance.

In summary, continuous measurement and refinement are essential for keeping AI systems effective over time. By implementing real-time monitoring systems, establishing feedback loops, and regularly retraining models, organizations can ensure that their AI solutions remain relevant, accurate, and aligned with their evolving business needs. This ongoing commitment to refinement enables AI to deliver sustained value and adaptability in dynamic business environments.

Example: Implementing Gen AI in a Financial Services Firm

To illustrate how organizations can apply the structured AI implementation process outlined in this chapter, this section presents a real-world example of a financial services firm integrating Gen AI into its customer support operations. This case study demonstrates each of the five steps—assessing business needs, running a pilot project, building the right infrastructure, scaling AI solutions, and continuously refining AI models—and highlights the tangible benefits of a strategic, well-executed AI deployment.

A midsized financial services firm identified the need to improve customer support as part of its broader digital transformation strategy. Customer inquiries were handled manually by support teams, leading to delays, inconsistent service quality, and rising operational costs. After assessing the business needs, the firm realized that implementing a Gen AI–powered chatbot could streamline the process, improve customer satisfaction, and reduce costs. The leadership team prioritized this initiative as a high-impact use case, given its potential to directly address customer pain points while aligning with strategic goals of improving operational efficiency and enhancing client experiences.

Step 1: Assessing Business Needs and Identifying High-Impact Use Cases   The firm's leadership began by analyzing internal data and customer service feedback to identify specific inefficiencies in the current support system. The team observed that a large portion of customer inquiries involved repetitive, simple queries—ideal for AI automation. Through cross-departmental workshops involving the customer service, IT, and operations teams, the company concluded that AI-driven customer support could significantly improve response times while freeing up human agents to handle more complex issues. With robust historical customer interaction data, the firm had the right foundation to support this AI-driven initiative.

Step 2: Pilot Projects: Testing AI Solutions in Controlled Environments   The firm initiated a pilot project by deploying an AI-powered chatbot to handle basic customer inquiries such as account balances, transaction history, and frequently asked questions (FAQs). The scope was intentionally narrow to allow close monitoring while still providing meaningful impact. The IT and customer support teams collaborated to integrate the chatbot into the company's website and mobile app, ensuring cross-functional input. Success metrics were clearly defined up front: reducing average response time to under one minute, improving customer satisfaction scores, and maintaining a high resolution rate for inquiries handled by the chatbot. Regular feedback was collected from customers and employees, allowing the team to identify areas for improvement and tweak the system as needed.

Step 3: Building the Right Infrastructure and Selecting Tools   To ensure scalability, the firm chose a cloud-based AI platform—Microsoft Azure AI—that allowed seamless integration with its existing CRM system. The platform provided prebuilt models and natural language processing (NLP) tools, accelerating the development process and allowing for future scalability. Data governance policies were also strengthened to ensure compliance with financial industry regulations, focusing on secure data storage and processing customer data in line with privacy laws.

Step 4: Scaling AI Solutions Across the Organization   After the pilot demonstrated positive results—such as a 60% reduction in customer response time and a 20% increase in customer satisfaction—the firm developed a scaling plan. The plan included rolling out the AI chatbot across all digital channels and extending its capabilities to handle more complex queries. SOPs were established to ensure consistency in how the AI solution was deployed and monitored across departments. Additionally, the firm launched comprehensive training programs for its customer service team, equipping them with the skills to collaborate effectively with the AI system and handle escalated cases.

Step 5: Continuous Measurement and Refinement   As the AI chatbot was scaled across the organization, continuous monitoring systems were put in place to track its performance in real time. Dashboards were set up to monitor key metrics like response times, accuracy rates, and customer satisfaction scores. Feedback loops were established, with regular input gathered from customers and employees. Based on this feedback, the AI models were retrained periodically to improve accuracy, especially in handling complex customer queries, ensuring that the system adapted to evolving business needs.

Outcome   Within six months of scaling the AI chatbot, the firm saw a 25% reduction in operational costs for its customer service department and a 15% increase in overall customer retention rates, as customers reported higher satisfaction with faster and more accurate responses. By implementing AI in a structured, measured way, the firm was able to leverage technology to significantly improve both customer experiences and operational efficiency while positioning itself for further AI-driven innovations across the organization.

This concrete example illustrates how a structured approach to AI implementation—starting with identifying high-impact use cases, running a controlled pilot, and scaling with the right infrastructure—can yield significant benefits.

Key Takeaways for Leadership

The practical steps for implementing Gen AI require a structured approach that starts with identifying high-impact use cases and progresses through piloting, infrastructure development, scaling, and continuous improvement. By following these steps, leaders can ensure that AI integration is successful and delivers sustained value.

Leadership should:

- **Align AI projects with strategic goals** to maximize impact
- **Involve cross-functional teams** to ensure collaboration and holistic AI adoption
- **Standardize processes and train employees** for seamless scaling
- **Establish monitoring and feedback systems** to refine AI solutions continuously

Case Studies: Success Stories of Gen AI in Action

Manufacturing has entered a new era driven by the convergence of data availability, computational power, and AI. Gen AI, in particular, has emerged as a transformative force, enabling manufacturers to optimize processes, improve quality, and accelerate innovation. Unlike traditional automation, which follows rigid, predefined rules, Gen AI learns from data, identifies patterns, and generates insights that empower decision-makers to adapt to dynamic conditions. This adaptability is especially valuable in manufacturing, where variability in raw materials, production environments, and market demand often disrupts operations and challenges efficiency.

Yet the successful adoption of Gen AI is not as simple as installing new software or automating a single process. Manufacturing leaders must move beyond the question of *what* AI can do and instead focus on *why* AI matters to their organization's broader strategy. The real challenge lies not in selecting an AI tool but in defining the *value* it creates. AI should be viewed not as a standalone technology deployment but as a catalyst for competitive differentiation, strategic agility, and long-term resilience. True leadership in AI adoption means asking, "Does this technology strengthen our ability to respond to supply chain volatility?

Does it enhance the autonomy of decision-making on the production floor? Does it reduce operational risk and drive sustainable profitability?"

This chapter presents case studies from manufacturing organizations of various sizes, industries, and levels of AI maturity. These real-world examples illustrate not only how AI has been applied to solve specific challenges but also how it has redefined the way these businesses operate, innovate, and compete. The companies highlighted here have successfully aligned AI adoption with strategic business goals by first understanding *why* they needed AI before determining *what* it should do.

By examining these success stories, leaders can gain insights into more than just AI's technical capabilities: they can learn how to evaluate AI investments through the lens of long-term business value. The organizations that have truly benefited from AI are those that have integrated it into their core decision-making frameworks, ensuring that AI adoption is not just a tool for automation but a driver of transformational change.

Case Study 1: Enhancing Quality Control in Electronics Manufacturing

Quality control is a critical aspect of manufacturing, especially in industries like electronics, where precision is paramount. Even minor defects in printed circuit board assemblies (PCBAs) can lead to significant issues, including costly rework and customer dissatisfaction. A European electronics manufacturer supplying PCBAs for automotive and aerospace applications faced challenges as increasing production volumes exposed flaws in its manual inspection processes. The company relied on human inspectors to identify soldering issues, component misalignments, and surface defects. Despite the team's expertise, variability in inspection results and the slow pace of manual checks led to rising defect rates and operational inefficiencies.

Recognizing the limitations of traditional automated optical inspection (AOI) systems—which often failed to detect subtle defects on complex, multilayer boards—the leadership team explored advanced solutions. After thorough research, including consultations with industry peers and pilot tests with various vendors, the company implemented an AI-driven visual inspection system. We can see how these compare in Figure 8-1.

This system utilized deep learning algorithms and high-resolution imaging to enhance defect detection capabilities. The implementation involved installing cameras at key inspection points along the production line and training the AI model using historical defect data. A phased approach allowed validation of the model's accuracy on a single production line before scaling up. Operators received training to interpret AI-generated reports and collaborate with data scientists to refine the model as needed.

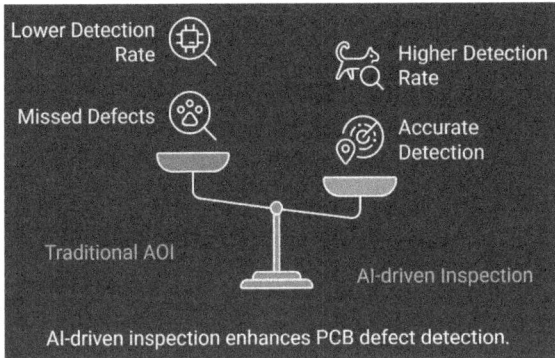

Figure 8-1: Before and after defect detection comparison

The results were remarkable. Within six months, the defect rate dropped significantly, leading to substantial savings in rework costs and warranty claims. Inspection throughput more than doubled, enabling comprehensive inspection coverage without increasing labor costs. Beyond these measurable outcomes, the AI system provided insights into the root causes of defects, such as identifying correlations between soldering defects and minor fluctuations in ambient humidity. This information prompted adjustments to the plant's climate control systems, further enhancing product quality. This process can be seen in Figure 8-2.

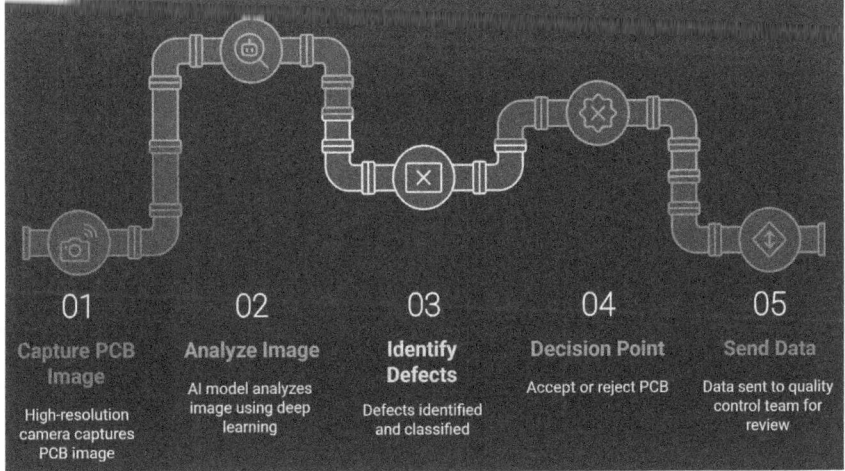

Figure 8-2: AI model decision pathway flowchart

From a leadership perspective, this case underscores the importance of selecting an AI model that not only delivers accurate results but also offers explainability. Transparency in AI decision-making was crucial for gaining the trust of quality control teams, who initially feared job displacement. By involving these teams in the AI implementation process and demonstrating how the technology augmented their work, the company fostered a culture of collaboration. This experience also highlights the value of starting with pilot projects, allowing for iterative improvements before full-scale deployment.

Source: Chung, Audrey, et al. "DVQI: A Multi-Task, Hardware-Integrated Artificial Intelligence System for Automated Visual Inspection in Electronics Manufacturing." arXiv.org (2023).

Case Study 2: Predictive Maintenance in Automotive Manufacturing

Maintenance is often viewed as a necessary expense in manufacturing, but when managed intelligently, it can become a source of competitive advantage. A United States-based automotive manufacturer experienced this transformation firsthand when it adopted AI-powered predictive maintenance to address chronic machine downtime. The company's engine production facility relied heavily on CNC machines to shape and finish critical components like cylinder heads and crankshafts. However, unplanned equipment failures disrupted production schedules, delayed customer deliveries, and inflated maintenance costs.

The maintenance team relied on time-based maintenance schedules, performing routine checks every few weeks regardless of the machines' actual condition. Although this approach prevented some failures, it also led to unnecessary maintenance activities and missed early signs of wear. The decision to implement predictive maintenance stemmed from a strategic initiative to improve overall equipment effectiveness (OEE). Senior executives tasked a cross-functional team to explore available technologies, considering factors like data integration capabilities, model accuracy, and vendor support.

After months of research and vendor demonstrations, the company partnered with Kortical, a firm known for its expertise in predictive analytics for manufacturing. Kortical's platform was chosen because of its ability to ingest diverse data streams—temperature readings, vibration patterns, and electrical current signatures—without requiring extensive sensor retrofitting. The AI model utilized advanced anomaly detection algorithms, comparing real-time data against historical patterns to predict potential failures. The process that Kortical was enabling was one of continuous improvement, as represented in Figure 8-3.

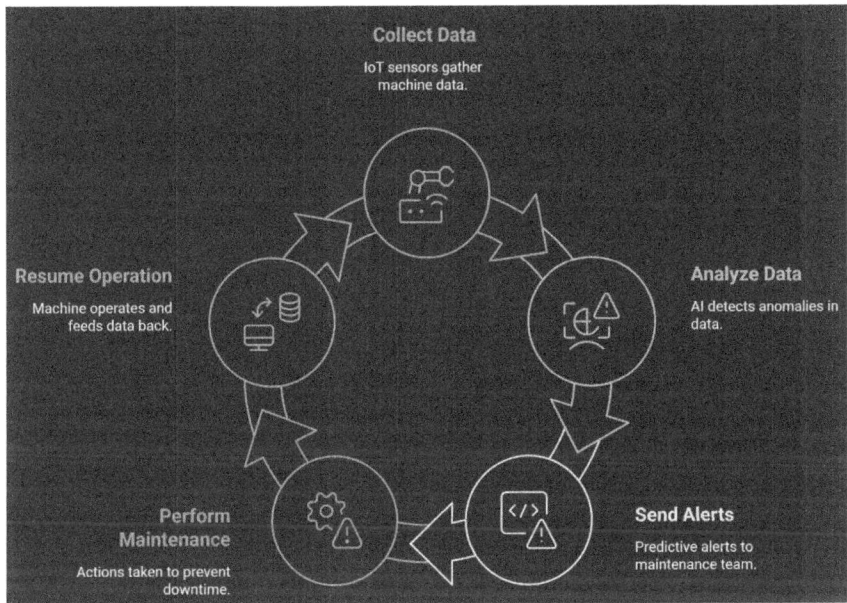

Figure 8-3: Predictive maintenance continuous improvement cycle powered by AI

The implementation process involved several stages. Initially, the company installed additional sensors on 100 critical machines and used historical maintenance records to train the AI model. The system's predictions were monitored for three months without intervention to assess their accuracy. During this trial, the model correctly identified 92% of the failures that occurred, exceeding the target accuracy rate set by the maintenance team. The platform was then fully integrated with the plant's computerized maintenance management system (CMMS), automating work-order generation based on AI predictions.

The impact was transformative. Unplanned downtime fell by 60%, translating into a monthly production increase equivalent to 200 additional engine units. Maintenance costs decreased by 25% due to reduced reliance on time-based activities and better-targeted interventions. An unexpected benefit was the improvement in technician morale; instead of reacting to breakdowns, teams could plan their work more effectively. Moreover, the data collected during the process helped identify systemic issues, such as insufficient lubrication in certain CNC machines, prompting long-term process improvements.

For manufacturing leaders, this case underscores the importance of choosing AI solutions capable of integrating with existing systems. The success of the predictive maintenance program was largely due to the seamless connection between Kortical's platform and the plant's CMMS. Additionally, the company benefited from engaging frontline maintenance staff early in the process, ensuring that the AI system addressed real-world operational needs rather than being a theoretical, top-down initiative.

Source: Kortical. (2023). "Predicting Failures Using Connected Vehicle Data: Insights from Ford Motor Company." `https://kortical.com/case-studies/ford-predicting-failures-ai-example`.

Case Study 3: AI-Augmented Product Design in Automotive Engineering

General Motors (GM), one of the world's largest automotive manufacturers, recognized that traditional product design methods were limiting its ability to achieve efficiency and innovation in vehicle engineering. The company sought to reduce vehicle weight and improve component performance but found that conventional CAD-based design approaches required excessive manual iteration, often constrained by engineers' preconceived notions and limited computational resources. Given the increasing importance of lightweighting in vehicle design to enhance fuel efficiency and meet regulatory standards, GM turned to Generative AI to radically transform its approach.

The company partnered with Autodesk to leverage its Fusion 360 Generative Design technology, a cloud-based AI-powered design platform that could autonomously generate and evaluate thousands of design alternatives based on specified parameters such as weight reduction, material strength, and manufacturability. The AI model analyzed structural constraints and simulated performance in real-world conditions, eliminating inefficient design elements while preserving functional integrity. One of the first applications was redesigning a seat bracket, a seemingly minor but structurally critical component responsible for securing the seat-belt fastener to the seat and floor. The results were transformative, as can be seen in Figure 8-4.

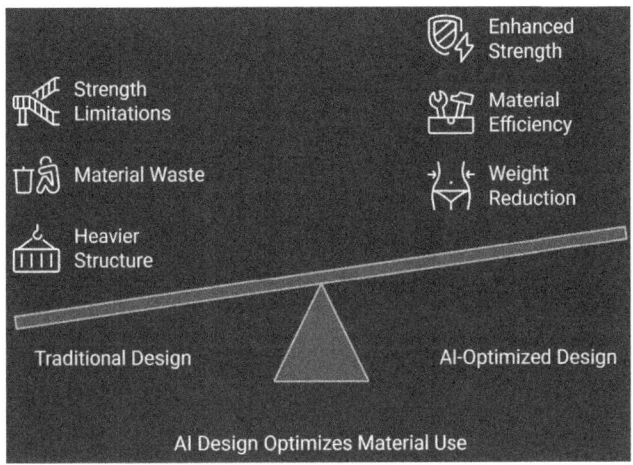

Figure 8-4: AI versus traditional CAD design comparison

The AI-generated seat bracket was 40% lighter and 20% stronger than its traditionally designed counterpart. This weight reduction contributed to improved fuel efficiency and vehicle safety while simultaneously reducing material costs. More importantly, the Generative AI process reduced design cycle time by 50%, accelerating time-to-market and allowing engineers to focus on refining AI-generated designs rather than manually drafting multiple iterations.

For manufacturing leaders, this case study underscores the importance of integrating AI driven design tools to enhance product innovation while maintaining cost and time efficiency. The key insight is that AI does not replace engineers but instead augments their capabilities, allowing them to explore a wider range of design solutions that human intuition alone may not have considered. However, leaders must recognize that successful AI adoption in design requires a cultural shift: engineers initially resisted AI-generated designs, questioning their feasibility. Only through iterative validation and real-world testing did the team gain confidence in the technology. This highlights a broader lesson: organizations must foster a culture of trust in AI-driven insights, ensuring that teams view AI as a collaborator rather than a competitor.

Source: Autodesk. (2021). "General Motors: Generative design in car manufacturing." `https://www.autodesk.com/customer-stories/general-motors-Generative-design`.

Case Study 4: AI-Driven Supply Chain Optimization in Beverage Production

Shanghai Pepsi-Cola Beverage Co. Ltd., a key subsidiary of PepsiCo, faced growing challenges in optimizing its production operations and supply chain planning. As a beverage manufacturer with a highly variable demand cycle, the company struggled to align production levels with fluctuating consumer preferences, raw material availability, and distributor constraints. Traditional supply chain management approaches relied on historical forecasting models, which often led to overproduction, excess inventory, or stockouts that impacted customer satisfaction. The company needed a dynamic, AI-driven approach that could optimize production in real time while accounting for multiple constraints simultaneously.

In response, Shanghai Pepsi-Cola implemented Blue Yonder's Production Planning AI platform, an advanced system capable of analyzing demand trends, supply constraints, and production capacity in real time. Unlike static forecasting methods, this AI solution continuously monitored internal and external variables, adjusting production schedules autonomously based on changing conditions. The system also integrated with the company's enterprise resource planning (ERP) software, allowing supply chain managers to make data-driven decisions on the fly.

It is important to note that this was more than just supplying decision-makers with a new dashboard; this was about using AI for enhanced decision-making for production. By leveraging a human interface, Blue Yonder is able to give those decision-makers insights into what and why changes are made.

The impact was significant. The AI-driven system reduced production waste by 15% and increased overall asset utilization, ensuring that machinery and labor resources were deployed efficiently. Additionally, real-time optimization enabled the company to respond 40% faster to demand fluctuations, minimizing disruptions in the supply chain. One of the most valuable outcomes was the ability to run "what-if" simulations, allowing decision-makers to test different scenarios—such as raw material shortages or seasonal demand spikes—before making costly adjustments.

For manufacturing leaders, this case study highlights the strategic advantage of leveraging AI to move from reactive to proactive supply chain planning. Many organizations still rely on static models that fail to capture real-time complexity. AI enables businesses to dynamically adjust operations, aligning production with demand while reducing waste and inefficiencies. However, leaders must ensure organizational alignment for AI adoption to succeed. Pepsi's success was partly due to early buy-in from both operational teams and executive leadership, ensuring that AI insights were trusted and acted upon. This underscores the need for leaders to prioritize cross-functional collaboration when deploying AI-driven supply chain solutions.

Source: Blue Yonder. (2024). "Shanghai Pepsi-Cola Beverage Co. Ltd. Selects Blue Yonder to Optimize Production Operations." `https://media.blueyonder.com/shanghai-pepsi-cola-beverage-co-ltd-selects-blue-yonder-to-optimize-production-operations.`

Case Study 5: AI-Enabled Energy Optimization in Glass Manufacturing

Saint-Gobain, a global leader in glass and building materials manufacturing, faced rising operational costs due to inefficient energy consumption across multiple production facilities in North America. With energy prices fluctuating and sustainability becoming a greater priority, the company needed a centralized approach to monitor and optimize energy use across its operations. Historically, each facility managed its own energy procurement and efficiency strategies, leading to inconsistent results and missed opportunities for cost savings.

To address these challenges, Saint-Gobain partnered with Schneider Electric to implement the EcoStruxure Resource Advisor, an AI-driven energy management platform capable of analyzing energy usage patterns across all sites in real time. The system aggregated data from hundreds of sensors embedded in production lines, furnaces, and HVAC systems, identifying inefficiencies such as prolonged furnace idling, suboptimal temperature controls, and peak electricity consumption hours. The AI continuously processed these insights and provided recommendations for adjustments, enabling operators to optimize energy use proactively. This cycle, as can be seen in Figure 8-5, offered Saint-Gobain a cycle of AI-driven continuous improvement. We cannot expect AI to be a magic bullet out of the box; deploying a system that supports a cycle of continuous improvement is driving key to value.

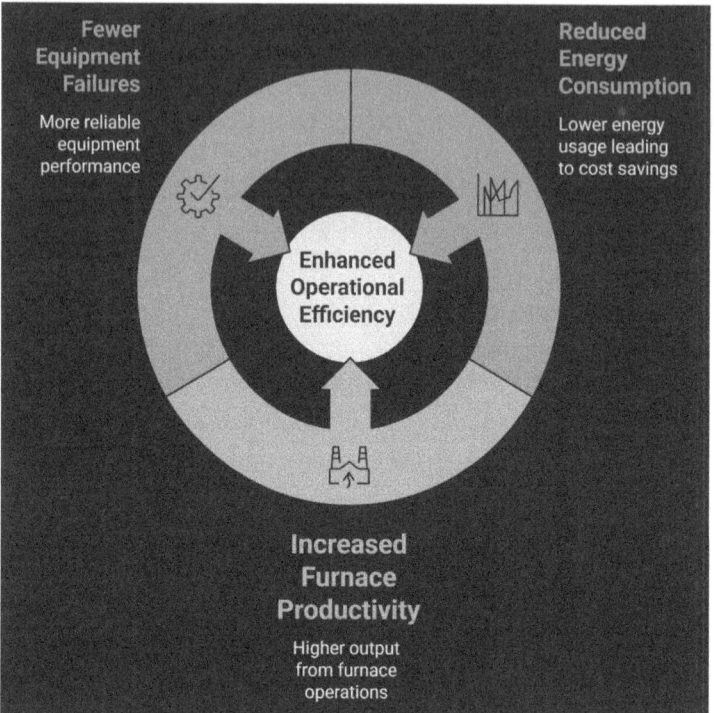

Figure 8-5: Energy savings and efficiency improvement cycle

The results were impressive. Over the course of 16 years, the company saved over $50 million in energy costs, demonstrating the long-term financial impact of AI-driven optimization. By centralizing energy procurement decisions, Saint-Gobain was able to negotiate better electricity rates and take advantage of demand-response programs. Additionally, the company reduced its carbon footprint, aligning with corporate sustainability goals and regulatory expectations.

For manufacturing leaders, this case presents a compelling example of how AI-driven energy optimization can transform a cost center into a value driver. Energy management is often seen as a technical concern rather than a strategic priority, but Saint-Gobain's success demonstrates that AI can yield both significant cost savings and sustainability benefits. However, leaders should recognize that scaling an AI-driven energy program requires organizational commitment. Saint-Gobain's success was due in part to its long-term approach, ensuring continuous improvements rather than expecting immediate results. This case underscores the value of viewing AI investments as strategic enablers rather than one-time implementations.

Source: Schneider Electric. (2019). "Saint-Gobain Builds on $50 Million in Energy Savings." `https://perspectives.se.com/mining-metals-minerals/saint-gobain-extends-strategic-sourcing-program`.

Agentic AI in Manufacturing: Successes, Challenges, and Failures

Agentic AI represents a significant evolution in manufacturing technology. Unlike traditional AI, which operates based on predefined rules and datasets, Agentic AI systems possess decision-making capabilities that allow them to adapt dynamically to new conditions, optimize processes in real time, and autonomously manage tasks without constant human oversight. In manufacturing, these capabilities translate into smarter, self-optimizing production lines, responsive supply chains, and adaptive maintenance systems.

However, with greater autonomy comes greater complexity. Although some companies have successfully leveraged Agentic AI to achieve remarkable operational improvements, others have encountered difficulties, ranging from technical challenges to organizational resistance. The following examples illustrate both sides of this journey.

Success Stories of Agentic AI in Manufacturing

Agentic AI is currently looked at as one of the latest iterations in disruptive AI technology. Some companies may claim to use Agentic AI, but often they are mislabeling machine learning or Generative AI. The following are case studies of Agentic AI moving toward mainstream deployments and adoption within organizations, but in small, niche areas. It is Agentic AI that will help companies truly move the needle in many critical business areas.

Case Study 1: Agentic AI for Real-Time Process Optimization in a Steel Mill

In the highly energy-intensive world of steel manufacturing, even small inefficiencies in furnace operations can lead to significant cost overruns and suboptimal product quality. As a leading global steel producer, BlueScope recognized the need to modernize its production processes by reducing variability in furnace performance and minimizing unplanned downtime. The company's legacy control systems relied on human operators to manually adjust furnace parameters based on pre-established guidelines, which often resulted in inconsistent performance, excessive energy consumption, and occasional defects in the final steel products.

Traditional maintenance approaches compounded these challenges. Furnace maintenance at BlueScope was typically scheduled based on fixed time intervals rather than actual equipment condition, leading to excessive preventive maintenance (which increased operational costs) and unexpected breakdowns (which resulted in lost production time). Given the fluctuating composition of raw materials and environmental conditions, an AI-driven solution was necessary to bring real-time adaptability and autonomous decision-making into the manufacturing process.

Implementing an AI-Powered Predictive Optimization System

To address these challenges, BlueScope partnered with Siemens to implement Senseye Predictive Maintenance, an Agentic AI system capable of continuously analyzing and optimizing furnace operations. Unlike traditional AI models that provide insights but still require human intervention, this Agentic AI system was designed to autonomously adjust furnace parameters, predict equipment failures, and optimize energy consumption in real time, significantly reducing reliance on human operators for minute-to-minute decision-making.

The AI model was trained using historical production data, metallurgical models, and real-time sensor readings, allowing it to learn the complex interplay between temperature, pressure, and chemical composition in steelmaking. The system integrated reinforcement learning algorithms, enabling it to adapt dynamically to fluctuating raw material properties and changing environmental conditions. This was a crucial advantage, as even minor variations in ore quality or ambient humidity could impact furnace efficiency and final product integrity.

Furthermore, the AI model continuously monitored thousands of sensor data points across the facility, processing vibration, heat flux, power consumption, and emission levels in real time. When it detected potential inefficiencies or anomalies, it automatically recalibrated parameters such as oxygen flow, furnace temperature, and heat distribution, ensuring consistent steel quality with minimal energy waste. Additionally, by integrating predictive maintenance capabilities, the system could forecast equipment failures weeks in advance, allowing the maintenance team to proactively schedule interventions before unexpected breakdowns occur. This cycle, as seen previously in Figure 8-3, is the true AI-driven continuous improvement cycle.

Results and Impact

The impact of BlueScope's Agentic AI implementation was significant. The company reported a 15% reduction in overall energy consumption, which translated into millions of dollars in cost savings annually. By automating real-time process adjustments, the AI system also improved furnace productivity by 20%, allowing BlueScope to produce higher volumes of steel without expanding its energy footprint.

Additionally, the introduction of predictive maintenance analytics reduced unplanned downtime by 30%, giving the company greater control over production schedules and reducing the risk of supply chain disruptions. One of the most unexpected benefits was the AI system's ability to identify previously unknown process inefficiencies, such as subtle inconsistencies in how raw materials interacted with furnace conditions. By continuously learning and adjusting, the AI system provided actionable insights that enhanced long-term process optimization beyond what human operators could achieve alone.

Leadership Insights: The Strategic Value of Agentic AI

For manufacturing leaders, this case underscores the transformative impact of Agentic AI in complex industrial environments. Although traditional automation and predictive analytics provide value, they still rely on human oversight and intervention. In contrast, Agentic AI enables real-time, autonomous decision-making, making it particularly well-suited for high-variability processes like steel production.

However, leaders considering Agentic AI must recognize that its success depends on more than just technological implementation—it requires organizational alignment and cultural adaptation. At BlueScope, initial resistance from furnace operators and maintenance teams was a key challenge. Many feared that AI-driven automation would diminish their roles or make their expertise obsolete. To address this, the company engaged frontline workers early in the deployment process, providing transparency on how the AI system functioned and emphasizing that it was a collaborative tool rather than a replacement for human expertise.

Additionally, the importance of domain expertise in AI model training cannot be overstated. Although the AI system was highly sophisticated, its success relied on input from experienced metallurgists and process engineers who helped fine-tune its decision-making framework. Leaders must ensure that AI deployments involve a cross-functional team that includes both technology specialists and industry experts to contextualize AI recommendations and align them with operational goals.

BlueScope's experience also highlights a broader strategic lesson: AI adoption should be approached as an ongoing optimization journey rather than a one-time deployment. The company continued refining its AI models, incorporating new data sources, sensor integrations, and real-world feedback from operators, ensuring that the system evolved alongside the manufacturing environment.

Looking Ahead: Scaling AI Across the Manufacturing Sector

The success of Agentic AI in steel production serves as a blueprint for other manufacturers looking to enhance process efficiency, energy management, and predictive maintenance. Industries such as cement production, chemical processing, and automotive manufacturing share similar challenges where process variability, equipment longevity, and energy efficiency are critical concerns.

As AI technology advances, we can expect even more autonomy in industrial decision-making, where AI systems not only optimize production processes but also collaborate with other AI-driven systems across the supply chain, making adjustments in response to market conditions, supplier variability, and customer demand forecasts.

For forward-thinking manufacturing leaders, the key takeaway is clear: Agentic AI is not just an operational improvement tool—it is a long-term strategic enabler that can redefine how industrial processes are optimized, sustained, and scaled for the future.

Source: Adapted from Siemens. "Generative Artificial Intelligence Takes Siemens' Predictive Maint ..." `Press.siemens.com`, 5 Feb. 2024.

Case Study 2: Agentic AI for Autonomous Inventory Management in Plastics Manufacturing

For manufacturers relying on raw materials such as polypropylene and polycarbonate, efficient inventory management is critical. Maintaining the right balance between having sufficient raw material stock and avoiding excessive inventory holding costs is a persistent challenge, especially in industries with fluctuating supply chain conditions. A midsized plastics manufacturer in North America found itself struggling with inventory inefficiencies that led to frequent stockouts and overstocking, disrupting production schedules and increasing operational costs.

Like many traditional manufacturers, the company had long depended on manual inventory tracking and fixed reorder thresholds, meaning that purchasing decisions were based on historical consumption patterns rather than real-time demand and supplier performance data. As a result, unexpected spikes in demand often led to stockouts, forcing production slowdowns or expensive last-minute raw material purchases. Conversely, periods of low demand resulted in excess raw material purchases that sat in storage, tying up capital and increasing waste. These inefficiencies rippled through the supply chain, causing missed customer delivery deadlines and higher procurement costs.

The leadership team recognized that the root of these issues lay in static inventory planning models that lacked adaptability. They sought a solution that would enable real-time, automated decision-making based on dynamic production needs, supplier lead times, and external disruptions. After evaluating multiple AI-driven inventory management systems, they selected an Agentic AI platform that could autonomously monitor raw material consumption, predict future demand fluctuations, and place optimized purchase orders with suppliers.

Implementing an Autonomous AI Inventory System

The company deployed an AI-powered inventory management system built on Kinaxis RapidResponse, a leading AI-driven supply chain planning platform. Unlike traditional inventory management software that relies on fixed rules and historical averages, this Agentic AI system continuously learns from real-time production data, demand forecasts, and supplier behavior patterns.

To ensure seamless integration with existing operations, the AI model was first trained using historical production and procurement data, allowing it to recognize recurring demand cycles, supplier lead times, and common disruptions. The system was then linked to real-time IoT sensors on production equipment that tracked material consumption rates and flagged deviations from expected usage levels. Additionally, the AI was integrated with the company's electronic data interchange

(EDI) system, enabling it to communicate directly with suppliers and confirm material availability before placing purchase orders.

What set this AI apart was its autonomous decision-making capability. Instead of requiring human intervention, the system could independently adjust reorder points and purchase quantities based on real-time production flow. If a supplier was experiencing delays, the AI would automatically recommend an alternative supplier or adjust production schedules to accommodate slower deliveries. The system also simulated "what-if" scenarios, allowing supply chain managers to test how different inventory strategies would impact costs and efficiency.

Results and Impact

Within the first six months of deployment, the AI-driven approach reduced raw material stockouts by 40% and decreased excess inventory levels by 30%. This provided an immediate financial benefit, as the company no longer had to rely on expensive emergency raw material purchases or pay excessive warehouse storage fees for surplus inventory. Production teams reported that operations ran more smoothly, as they were no longer forced to halt production due to missing raw materials.

Beyond these measurable improvements, the AI brought a new level of supply chain resilience. Previously, procurement teams reacted to disruptions only after they occurred; but with real-time predictive analytics, they could anticipate supplier delays and adjust inventory strategies proactively. The AI also revealed previously unnoticed inefficiencies in supplier performance, allowing the company to renegotiate contracts and prioritize partnerships with more reliable raw material suppliers.

One unexpected benefit of the AI-driven system was its ability to improve sustainability efforts. By optimizing material usage and reducing excess stock, the company lowered its waste output and decreased its carbon footprint by minimizing unnecessary shipping and storage emissions. This helped the company align with its sustainability goals and improve regulatory compliance.

Leadership Insights: AI as a Strategic Supply Chain Enabler

For manufacturing leaders, this case demonstrates the power of AI-driven decision-making in inventory management. Many organizations still rely on manual tracking systems and reactive purchasing strategies, which can lead to supply chain inefficiencies, higher costs, and increased operational risks. Agentic AI provides a scalable solution that enhances supply chain resilience by enabling proactive, data-driven inventory planning.

However, successful implementation of AI-driven inventory management requires more than just technology investment. It demands a fundamental shift in how procurement and production teams interact with supply chain data. Initially, some employees resisted the new AI-driven approach, fearing it would replace human decision-making. To address these concerns, the company prioritized workforce training, demonstrating that the AI was not eliminating jobs but enhancing operational efficiency by reducing the time spent on repetitive tasks like purchase order management.

Additionally, data integration across supply chain partners is critical. AI models are only as good as the data they receive, so ensuring that suppliers provide accurate and timely information is essential. The company worked closely with suppliers to establish standardized data-sharing protocols, ensuring that the AI system had access to real-time lead times, pricing fluctuations, and shipment statuses.

Another key takeaway from this case is that AI-driven inventory systems must be adaptable to changing business conditions. As market dynamics shift, AI models must be continuously trained with updated data to maintain their accuracy and relevance. The company recognized this early on and established a feedback loop where procurement teams regularly reviewed AI recommendations, provided manual adjustments when necessary, and collaborated with data scientists to refine the model's predictive capabilities over time.

Looking Ahead: Expanding AI-Driven Inventory Management

Following the success of the AI-driven inventory system in managing raw materials, the company is now exploring expanding AI automation into other areas of supply chain management, such as demand forecasting for finished goods and predictive logistics planning. By integrating AI-driven decision-making into end-to-end supply chain processes, the company aims to further improve efficiency, reduce waste, and strengthen its ability to navigate supply chain disruptions.

This case serves as a blueprint for manufacturers looking to adopt AI in supply chain operations. Whereas traditional inventory management relies on static rules and human intuition, AI introduces autonomous, data-driven adaptability, making manufacturing more resilient, cost-efficient, and scalable. For leaders seeking to improve inventory control, reduce costs, and optimize supplier relationships, investing in Agentic AI is not just an operational enhancement—it is a long-term strategic advantage.

Source: Adapted from Fredrik Filipsson. "AI in Inventory Management for Manufacturing." Redress Compliance - Just Another WordPress Site, 5 June 2024.

Case Study 3: Autonomous Production Scheduling in a Textile Factory

The textile industry is highly dynamic, with fluctuating customer demand, rapid shifts in fashion trends, and tight production deadlines. Efficient production scheduling is essential for manufacturers to maintain on-time delivery, maximize machine utilization, and minimize costs. However, many textile factories still rely on manual scheduling methods, where planners adjust production schedules based on historical patterns rather than real-time operational data. This often leads to inefficiencies, delayed orders, and excessive downtime due to a lack of visibility into machine performance and order priority shifts.

A textile manufacturer in Southeast Asia was facing persistent production inefficiencies due to last-minute order changes, unexpected machine breakdowns, and an inability to dynamically adjust schedules. Schedulers manually updated production plans using spreadsheets and past production trends, failing to account for real-time machine status, material availability, and urgent customer demands. As a result, production throughput was inconsistent, with frequent bottlenecks and missed delivery deadlines, affecting customer satisfaction.

Implementing an AI-Powered Production Scheduling System

To modernize its scheduling process, the manufacturer implemented Rockwell Automation's FactoryTalk Analytics LogixAI, an Agentic AI solution designed to optimize industrial operations in real time. Unlike traditional scheduling systems that require manual intervention, this AI-driven solution autonomously generates and updates production plans based on machine performance, maintenance schedules, order priority, and material availability.

The implementation involved integrating IoT sensors and machine control systems across the factory floor, allowing the AI to continuously collect data on machine utilization, energy consumption, production output, and real-time order requirements. This data was then processed using predictive analytics and reinforcement learning algorithms, enabling the AI system to anticipate bottlenecks and dynamically adjust production schedules before disruptions occurred.

One of the most critical advantages of this system was its ability to autonomously reallocate tasks. If a high-priority order was introduced, the AI could automatically adjust the production sequence, ensuring that the urgent order was completed without disrupting other scheduled tasks. Similarly, if a machine failure occurred, the system could redistribute workloads to other available machines, preventing bottlenecks and ensuring continuous operations.

Results and Impact

The AI-driven scheduling system delivered measurable improvements within six months. Production throughput increased by 18% as machine utilization became more efficient and downtime was minimized. Schedule adherence improved from 65% to 92%, meaning that nearly all production deadlines were met on time. Additionally, the system identified previously unnoticed inefficiencies, such as certain machines experiencing a disproportionate number of breakdowns. With this insight, the manufacturer was able to implement targeted predictive maintenance strategies, further reducing downtime and extending machine lifespan.

Beyond these quantitative improvements, the AI system introduced a new level of production agility. The factory could now respond to sudden demand fluctuations faster and adjust production schedules dynamically based on incoming orders. This increased flexibility gave the company a competitive advantage, enabling it to take on more last-minute, high-value contracts that competitors relying on traditional scheduling methods could not handle.

Leadership Insights: How AI-Driven Scheduling Transforms Operations

For manufacturing leaders, this case highlights the strategic advantage of Agentic AI in production scheduling. Many companies still rely on human planners using static models, which fail to adapt to real-time operational changes. By implementing AI-driven scheduling systems, manufacturers can optimize resource utilization, minimize downtime, and increase agility—all of which directly impact profitability and customer satisfaction.

However, successful adoption of AI-driven scheduling requires high-quality data and seamless integration with factory floor systems. In this case, the manufacturer invested in IoT sensors and machine connectivity infrastructure to ensure that the AI had access to accurate, real-time information. Without this data, the AI's scheduling recommendations would not have been as effective. Leaders should prioritize modernizing data collection systems before deploying AI-driven scheduling solutions.

Another key takeaway is that AI-driven production scheduling is not just a technological upgrade—it is a cultural shift. Initially, production planners were skeptical about AI making scheduling decisions, fearing that it would replace human judgment. To address these concerns, the company provided comprehensive training to planners and factory supervisors, demonstrating how the AI system worked as a decision-support tool rather than a replacement. Once teams saw the benefits of reduced scheduling complexity and fewer disruptions, they became strong advocates for the system.

This case also underscores the need for continuous AI learning and adaptation. Unlike traditional scheduling software that follows pre-defined rules, Agentic AI continuously improves its decision-making by learning from past scheduling outcomes. The manufacturer set up a feedback loop where human operators reviewed AI-generated schedules and provided insights on unexpected disruptions. This iterative approach ensured that the AI system became more accurate and responsive over time.

Expanding AI-Driven Scheduling Across the Industry

Following the success of AI-driven production scheduling in this textile facility, the company is now exploring scaling AI adoption to other areas of its operations. Plans are underway to integrate AI with supply chain logistics, inventory management, and workforce scheduling to create a fully autonomous and interconnected manufacturing ecosystem.

This case study serves as a model for manufacturers in fast-paced, high-variability industries looking to implement AI-driven scheduling solutions. By reducing reliance on manual decision-making, enabling real-time schedule adjustments, and improving overall efficiency, Agentic AI is reshaping how manufacturers approach production planning.

For leaders considering AI-driven scheduling, the key lessons are as follows:

- Data infrastructure must be modernized before AI implementation. Real-time production data is essential for accurate scheduling.

- Cross-functional collaboration is critical. Engage production planners and factory workers early in the AI deployment process to build trust and adoption.

- AI-driven scheduling is an evolving process. Set up continuous feedback loops to refine AI decision-making over time.

- Scalability is key. Successful AI scheduling can be expanded to other operational areas, including inventory management, workforce planning, and supply chain logistics.

Looking Ahead: AI as a Competitive Differentiator in Manufacturing

As AI continues to advance, Agentic AI systems will become the standard for production scheduling in complex manufacturing environments. Textile manufacturers, in particular, stand to benefit from AI's ability to handle last-minute order changes, optimize machine usage, and ensure on-time production.

Manufacturing leaders who invest in Agentic AI for scheduling today will be the ones who set industry benchmarks for efficiency, agility, and customer responsiveness in the future. Companies that hesitate risk being outperformed by competitors that leverage AI's ability to make real-time, intelligent scheduling decisions.

This case provides a clear roadmap for AI adoption, showing how real-time data, autonomous decision-making, and continuous learning can transform production planning from a reactive process into a proactive, value-generating function. The lesson for leaders is clear: AI is not just a tool for automation—it is a strategic asset that can drive long-term manufacturing excellence.

Source: Adapted from FactoryTalk Analytics LogixAI | FactoryTalk." Rockwell Automation, www.rockwellautomation.com/en-us/products/ software/factorytalk/operationsuite/analytics-logixai.html.

Initial Challenges and Lessons Learned

Although many companies experience success with Agentic AI, initial deployment often presents challenges that manufacturing leaders must address. Here are some recurring obstacles observed in early-stage implementations:

- **Data silos:** Manufacturing data is often scattered across legacy systems, making it difficult for Agentic AI models to access the comprehensive datasets they need to learn effectively.

- **Cultural resistance:** Operators and middle managers sometimes perceive AI as a threat to job security, leading to subtle resistance during implementation.

- **Model generalization issues:** AI models trained on historical data may struggle to adapt to entirely new conditions, requiring ongoing model refinement and retraining.

- **Integration complexity:** Connecting AI platforms with existing manufacturing execution systems (MES) and ERP software can be more time-consuming than anticipated.

Despite these challenges, companies that succeed typically adopt a phased implementation strategy, starting with pilot projects to test and refine the technology before scaling it across multiple production lines.

Not all AI projects in manufacturing succeed. The following case studies illustrate how certain deployments failed, providing valuable lessons for leaders to avoid similar pitfalls.

Failure Case 1: AI-Driven Maintenance System in a Heavy Equipment Manufacturer

A large manufacturer of construction equipment invested heavily in an AI-driven predictive maintenance system, expecting to reduce machine downtime and improve maintenance efficiency. The AI system, developed internally using open-source machine learning frameworks, was tasked with predicting component failures across a fleet of CNC machines, hydraulic presses, and robotic welders.

What Went Wrong

The project failed due to poor data quality and insufficient domain expertise during model development. Sensor data was inconsistent, with many sensors either malfunctioning or transmitting inaccurate readings. Additionally, the AI model was trained on limited historical data that did not account for the variability in equipment usage patterns. Maintenance teams, skeptical of the model's predictions, continued following their existing routines, rendering the AI recommendations irrelevant.

Key Lesson for Leaders

Data quality is the foundation of reliable AI performance. Before deploying Agentic AI for predictive maintenance, manufacturing leaders must conduct comprehensive data audits and involve maintenance experts to contextualize the model's outputs. Additionally, establishing clear accountability for data maintenance ensures that sensors and logging systems remain reliable over time.

Failure Case 2: Autonomous Material Handling System in a Warehouse

A United States-based automotive parts supplier deployed an AI-driven autonomous guided vehicle (AGV) system to automate material transport within its main distribution warehouse. The system used reinforcement learning algorithms to optimize routes and minimize transit times. However, the project was suspended after six months due to operational disruptions.

What Went Wrong

The AI agents were trained under controlled conditions that did not reflect the variability of real-world warehouse operations. For instance, the system struggled to adapt when new racking configurations were introduced, causing AGVs to collide with obstacles or stall in dead-end pathways. Moreover, warehouse staff were not adequately trained to troubleshoot the AI's decisions, leading to extended downtime during incidents.

Key Lesson for Leaders

AI models must be trained on datasets that accurately represent the full range of operating conditions. Additionally, investing in staff training and change management initiatives is essential to ensure that human operators can collaborate effectively with Agentic AI systems.

Insights for Manufacturing Leaders

The successful integration of AI in manufacturing is not just about adopting new technology: it is about aligning AI capabilities with business objectives, ensuring organizational buy-in, and building an AI-ready infrastructure. As demonstrated by the case studies, manufacturing leaders must go beyond understanding the use case and instead focus on the value case—the tangible and strategic benefits AI can provide over the long term. The following insights serve as guiding principles for leaders considering AI adoption in their organizations:

1. **AI is a strategic investment, not just an operational tool**
   Many organizations approach AI as a solution for incremental automation, targeting specific inefficiencies in quality control, maintenance, or scheduling. However, companies that reap the greatest benefits from AI treat it as a long-term strategic investment that transforms how decisions are made, how supply chains function, and how value is created. AI should be viewed not as an isolated tool but as a central component of a data-driven, agile, and scalable manufacturing ecosystem.

2. **Data infrastructure is the foundation of AI success**
   The most sophisticated AI models are only as good as the data they receive. Companies that struggle with fragmented data sources, poor data quality, or legacy IT systems will find AI implementations failing to deliver expected results. Before deploying AI, manufacturers must ensure they have real-time, high-quality data collection systems in place. This may require investments in IoT sensors, ERP integrations, and cloud-based analytics platforms to create a unified, structured data environment.

3. **Cross-functional collaboration is critical**
   AI adoption is not solely an IT or operations initiative; it requires collaboration across departments, including engineering, production, procurement, and supply chain management. Many AI projects fail because they are implemented in silos without considering how AI-driven decisions will be operationalized across the organization. Successful AI deployments involve multidisciplinary teams that bring together data scientists, process engineers, factory operators, and executives to ensure that AI insights are practical and actionable.

4. **AI must be explainable and trustworthy**
   In industries like manufacturing, where precision and reliability are critical, AI models must provide more than just accurate predictions: they must also be transparent and explainable. Workers and decision-makers need to understand why an AI model made a specific recommendation, to trust and act on its insights. Companies should prioritize AI solutions with built-in explainability features, allowing human operators to validate AI-driven decisions and refine models based on operational experience.

5. **Change management and workforce engagement are essential**
   One of the biggest challenges in AI adoption is overcoming employee resistance and fear of job displacement. Many workers view AI as a threat rather than an enabler, leading to skepticism or even rejection of AI recommendations. Companies that successfully integrate AI prioritize change management initiatives, clearly communicating how AI enhances human decision-making rather than replacing it. Training programs should be implemented to ensure that factory operators and maintenance teams are equipped to work alongside AI systems effectively.

6. **Start with pilot projects and scale gradually**
   Attempting to implement AI across an entire operation at once is a common mistake. A more effective approach is to start small with pilot projects, assess their impact, refine the technology, and then scale incrementally. Pilot programs help identify technical challenges, process gaps, and organizational resistance before broader deployment. Leaders should establish clear key performance indicators (KPIs) for each pilot to measure success and gather insights for larger rollouts.

7. **AI-driven decision-making must be continuously optimized**
   AI models are not static: they must evolve as new data becomes available and business conditions change. Successful companies invest in continuous learning and model retraining, ensuring that AI systems remain relevant and effective over time. This requires a structured feedback loop where human operators and AI systems work together to refine decision-making processes. Organizations that treat AI as an ongoing optimization journey rather than a one-time deployment achieve greater long-term value.

8. **AI deployment is not just about cost savings—it's about competitive advantage**
   Although cost reduction is often a key driver for AI adoption, the real value of AI lies in its ability to enhance agility, resilience, and

innovation. Companies that effectively leverage AI gain a competitive edge by being able to predict disruptions before they happen, respond faster to market changes, and optimize operations dynamically. AI-driven businesses outperform competitors that still rely on static, manual decision-making processes.

Conclusion: Restating the Case for AI in Manufacturing

Manufacturing is undergoing a profound transformation, driven by AI's ability to process vast amounts of data; make real-time, autonomous decisions; and continuously improve operations. From quality control and predictive maintenance to inventory optimization and production scheduling, AI has proven to be more than just an automation tool—it is a strategic enabler of resilience, efficiency, and long-term profitability.

However, the journey toward AI-driven manufacturing excellence requires more than just technological adoption. Leaders must first define the value AI creates for their business, ensuring alignment with broader strategic objectives rather than focusing solely on the technology itself. Companies that ask "Why AI?" before "What AI?" are more likely to unlock transformational benefits that extend beyond immediate cost savings.

Moreover, successful AI adoption depends on organizational readiness. Investing in data infrastructure, workforce training, and cross-functional collaboration is just as critical as selecting the right AI model. The most forward-thinking manufacturers view AI not as a one-time investment but as an ongoing competitive differentiator, continuously refining their systems, processes, and decision-making frameworks as AI capabilities evolve.

The companies highlighted in this chapter provide a blueprint for success, demonstrating that AI is not just about automation—it's about creating smarter, more adaptive, and future-proof manufacturing ecosystems. As AI continues to advance, the leaders who embrace its potential, integrate it into core decision-making frameworks, and foster an AI-ready workforce will be the ones who define the future of global manufacturing.

The Future of Gen AI: Trends and Innovations to Watch

Thus far, this book has taken a structured playbook approach to implementing Gen AI, guiding organizations through a step-by-step process to integrate AI into their operations successfully. However, before moving further into scaling strategies and long-term AI adoption, it is important to take a brief step back to examine emerging trends and future innovations in the Gen AI space.

AI is evolving rapidly, with new advancements reshaping industries and unlocking transformational opportunities that extend beyond current best practices. Staying ahead of these trends is essential for leaders who want to maintain a competitive edge and prepare their organizations for the next wave of AI-driven transformation.

- This chapter provides a break from the tactical playbook approach to explore key trends, innovations, and challenges in the AI landscape. It covers

- Advancements in AI models and capabilities that are pushing the boundaries of AI's potential

- Emerging applications across industries, from AI-driven product design to healthcare innovations

- AI-enhanced decision-making and automation
- Ethical, regulatory, and governance challenges that organizations must address as AI adoption expands
- Emerging technologies like Quantum AI that could define the future of AI's capabilities
- AI-driven transformation in manufacturing
- Strategic considerations for leadership

By understanding these trends, leaders can make informed decisions, anticipate shifts in the AI landscape, and position their organizations for long-term success.

Advancements in AI Models and Capabilities

The field of AI is advancing rapidly, with new models and capabilities emerging at an unprecedented pace. The evolution of more sophisticated Gen AI models is expanding the technology's potential, allowing AI systems to perform tasks with greater accuracy, creativity, and adaptability.

One of the most significant advancements is the development of larger and more powerful AI models. Neural networks are becoming increasingly complex, enabling AI to process and generate highly nuanced content. The next generation of Gen AI models is expected to surpass current capabilities, producing sophisticated outputs across various forms of media, including text, images, and even interactive experiences.

Another breakthrough is the rise of multimodal AI systems. Future AI models will not be limited to processing a single type of data but will integrate multiple formats, such as text, images, and audio, within a single system. This will enable more versatile applications, such as virtual assistants that can simultaneously understand voice commands, display relevant visual information, and provide contextual responses in real time. The ability to seamlessly switch between different data modalities will make AI interactions more natural and intuitive.

The advancement of few-shot and zero-shot learning is also transforming AI capabilities. Traditionally, AI models required vast amounts of training data to perform effectively. However, recent developments allow AI to complete complex tasks with minimal data input. Few-shot learning enables AI to learn from a small set of examples, and zero-shot learning allows it to generalize and make accurate predictions without prior exposure to specific training data. These capabilities significantly reduce the need for extensive datasets, making AI deployment more efficient and adaptable across different industries and applications.

As AI models continue to grow in complexity and capability, these advancements will shape the future of AI-driven solutions, making them more powerful, flexible, and accessible across a wide range of real-world applications.

Emerging Applications Across Industries

As Gen AI capabilities continue to expand, new applications are emerging across a wide range of industries. These innovations are not only enhancing operational efficiency but are also transforming customer experiences, product development, and creative content generation. AI is increasingly being integrated into business processes, enabling organizations to optimize workflows, accelerate innovation, and personalize services in ways that were previously unattainable.

One of the most impactful developments is AI-driven product design, which is revolutionizing industries such as manufacturing, automotive, and fashion. AI systems can generate design variations based on specific parameters, including material constraints, functional requirements, and aesthetic preferences. This capability accelerates the development process by quickly iterating through potential solutions, reducing costs, and allowing for more personalized products. By streamlining ideation and prototyping, AI-driven design tools are enabling companies to bring new products to market faster and more efficiently.

In healthcare, AI is expected to play a crucial role in advancing medical diagnostics, personalized treatment plans, and drug discovery. Gen AI can analyze vast amounts of complex medical data to identify patterns that might be overlooked by human practitioners. By recognizing trends in patient symptoms, medical history, and genetic markers, AI can assist in diagnosing conditions earlier and suggesting tailored treatment options. Additionally, AI is being utilized to accelerate drug discovery, helping researchers identify promising compounds and predict their effectiveness, ultimately reducing the time and cost required to develop new treatments.

Creative industries are also experiencing a transformation with the integration of AI. From music composition to digital art and film production, Gen AI is enabling the rapid generation of original content. Artists and designers are increasingly leveraging AI as a collaborative tool, using it to generate ideas, refine concepts, and push creative boundaries.

AI-generated suggestions serve as inspiration, helping creators explore new styles and techniques while maintaining a human touch in the final product. As AI becomes more advanced, it is expected to further blur the lines between human and machine-generated creativity, opening new possibilities for artistic expression.

These emerging applications demonstrate how Gen AI is reshaping industries by automating processes, enhancing decision-making, and fostering innovation. As AI continues to evolve, its impact will become even more profound, driving efficiency, creativity, and personalized solutions across various sectors.

Enhanced Personalization and Customer Engagement

Gen AI's ability to process and analyze vast datasets is revolutionizing personalization in customer interactions. As AI systems become more advanced, they are enabling businesses to deliver highly customized experiences that adapt in real time to individual preferences and behaviors. This level of personalization is transforming how companies engage with their customers, making interactions more intuitive, relevant, and responsive.

One of the most significant advancements in this space is real-time personalization. AI-driven systems can now tailor experiences dynamically, adjusting recommendations, content, and interactions based on a customer's current context and past behaviors. This allows businesses to create personalized shopping experiences, curated content feeds, and adaptive service offerings that evolve with customer needs. By continuously analyzing user data, AI can refine its understanding of preferences, ensuring that each interaction feels increasingly relevant over time.

Conversational AI and virtual assistants are also becoming more sophisticated, offering more natural and effective interactions. Gen AI is enhancing virtual assistants with multimodal capabilities, enabling them to understand and respond using a combination of voice, text, and visual elements. This evolution is making AI-driven customer support more seamless, allowing businesses to deliver richer, more engaging experiences. Whether assisting with transactions, troubleshooting issues, or providing personalized recommendations, virtual assistants are reducing the need for human intervention while maintaining high-quality service.

Another key development is the use of AI for predictive customer insights. Businesses are leveraging AI to anticipate customer needs and proactively offer solutions, recommendations, and support before issues arise. By integrating predictive analytics into customer service, marketing, and sales functions, organizations can enhance customer satisfaction and loyalty. AI can detect behavioral patterns that signal potential concerns, allowing companies to address them in advance, improve retention, and strengthen brand relationships.

These advancements in AI-driven personalization are reshaping the way businesses interact with their customers. By delivering more relevant, timely, and proactive experiences, AI is not only enhancing engagement but also setting new standards for customer expectations in the digital age.

AI-Enhanced Decision-Making and Automation

Gen AI is playing an increasingly pivotal role in decision-making processes across industries, providing executives with data-driven insights that improve strategic planning and operational efficiency. As AI systems evolve, their ability to automate complex tasks is expanding, enabling organizations to optimize workflows, enhance productivity, and reduce costs. The integration of AI into business operations is not only streamlining routine processes but also transforming how companies approach long-term planning and execution.

One of the most impactful advancements in this area is AI-augmented analytics. AI-powered analytics tools are enhancing traditional data analysis by identifying trends, forecasting outcomes, and generating strategic recommendations based on real-time data. These AI-driven insights empower business leaders to make more informed decisions, reducing reliance on intuition and ensuring that strategies are backed by comprehensive, data-driven evaluations. From financial forecasting to market trend analysis, AI-augmented analytics is helping organizations stay agile and competitive in an increasingly complex business landscape.

The rise of autonomous operations is also reshaping industries by enabling AI-driven systems to manage entire processes without human intervention. In logistics, AI is optimizing supply chains by dynamically adjusting inventory levels and predicting delivery delays. In manufacturing, AI-driven automation is enhancing predictive maintenance, monitoring equipment performance, and preventing costly downtimes. Similarly, the energy sector is leveraging AI to regulate power distribution, monitor grid stability, and enhance resource efficiency. By taking over critical operational tasks, AI is allowing businesses to focus on innovation and strategic growth.

Another significant advancement is the adoption of intelligent process automation (IPA), which integrates AI with robotic process automation

(RPA) to streamline workflows and minimize manual intervention. AI systems are now capable of analyzing vast amounts of unstructured data, making intelligent decisions, and even initiating actions in real time. This enables businesses to automate end-to-end processes, from handling customer inquiries to processing financial transactions. By reducing human workload and improving process efficiency, IPA is helping organizations achieve greater scalability and operational resilience.

As AI continues to enhance decision-making and automation, businesses are reaping the benefits of improved efficiency, reduced costs, and more strategic resource allocation. The increasing sophistication of AI-driven analytics, autonomous operations, and intelligent automation is fundamentally changing how organizations operate, paving the way for more adaptive, intelligent, and future-ready enterprises.

Ethical and Regulatory Challenges in AI Implementation

As AI systems become more advanced and integrated into critical business functions, they introduce significant ethical and regulatory challenges that organizations must address. Ensuring responsible AI use requires proactive management of issues such as bias, data privacy, and transparency while also navigating an evolving regulatory landscape. Companies deploying AI must establish governance frameworks that mitigate risks and ensure compliance with legal and ethical standards, particularly as AI becomes more autonomous and influential in decision-making processes.

One of the most pressing challenges in AI ethics is bias and fairness. AI models learn from historical data, which may contain biases that lead to unfair or discriminatory outcomes. If these biases are not identified and corrected, AI systems can reinforce societal inequalities, impacting hiring decisions, credit approvals, medical diagnoses, and more. To address this issue, organizations must implement rigorous bias-detection and -mitigation strategies, ensuring that AI models promote fairness and inclusivity. Conducting regular audits, diversifying training data, and refining model parameters are crucial steps in reducing bias and maintaining ethical AI practices.

Data privacy and security also pose significant concerns in AI implementation. Many AI systems require access to vast amounts of sensitive data, raising issues related to confidentiality, user consent, and protection against data breaches. Regulations such as the General Data Protection Regulation (GDPR) and the California Consumer Privacy Act (CCPA) impose strict requirements on how organizations collect, store, and process data. Businesses must establish robust data governance policies that ensure compliance with these regulations while safeguarding consumer trust. This includes implementing encryption, anonymization techniques, and strict access controls to prevent unauthorized use or exposure of sensitive information.

Another major challenge is ensuring transparency and accountability in AI decision-making. As AI systems become more autonomous, it is crucial that their decision-making processes are explainable and auditable. Many AI models function as "black boxes," producing outcomes without clear explanations of how decisions were reached. Explainable AI (XAI) solutions are gaining prominence as they provide insights into AI decision-making, allowing users to understand and validate AI-generated outcomes. Transparency not only fosters trust among stakeholders but also helps organizations comply with regulatory requirements, particularly in high-stakes industries such as finance, healthcare, and legal services.

To ensure responsible AI use, organizations must take deliberate actions to establish governance frameworks that address these challenges. Conducting regular bias audits is essential to detecting and correcting unfair AI behavior. Companies should develop clear ethical guidelines that prioritize fairness, inclusivity, and accountability throughout AI model development and deployment. Strengthening data governance practices is another critical step. Organizations must ensure that AI systems handle personal data in compliance with privacy laws by implementing strict data access policies and continuous monitoring of data security practices.

Investing in XAI is also a crucial strategy for enhancing accountability. By adopting AI models that provide transparent decision-making insights, businesses can improve regulatory compliance and increase stakeholder confidence in AI-driven outcomes. XAI solutions help organizations meet ethical standards by ensuring that AI-generated decisions are interpretable, justifiable, and aligned with organizational values.

As AI continues to advance, ethical and regulatory considerations will remain central to its responsible deployment. Organizations that proactively address these challenges will not only mitigate risks but also build trust with customers, regulators, and society at large. Establishing strong governance frameworks, prioritizing fairness and transparency, and ensuring robust data protection measures will be key to fostering ethical AI innovation while complying with evolving regulatory requirements.

Preparing for Quantum AI and Other Emerging Technologies

Quantum computing, although still in its early stages, has the potential to revolutionize AI by enabling vastly more complex calculations at unprecedented speeds. The integration of quantum computing with artificial intelligence could unlock new possibilities for data processing, optimization, and machine learning, allowing organizations to tackle challenges that are currently beyond the reach of classical AI. Quantum AI remains an emerging field, but its potential to transform industries is becoming increasingly clear.

One of the most significant advancements in this space is the rapid development of quantum computing. Unlike traditional computers that process data using binary bits (0s and 1s), quantum computers leverage qubits, which can exist in multiple states simultaneously due to the principles of superposition and entanglement. This capability allows quantum computers to perform calculations exponentially faster than classical systems, leading to breakthroughs in fields such as cryptography, materials science, and complex simulations. As quantum hardware improves, its applications in AI will become more sophisticated, enabling faster and more efficient model training, enhanced data analysis, and more advanced neural network architectures.

The synergy between quantum computing and AI is expected to reshape industries by introducing new methods for problem-solving and optimization. Quantum AI, which combines quantum computing with machine learning, has the potential to solve problems that are currently intractable due to computational limitations. In pharmaceuticals, Quantum AI could accelerate drug discovery by simulating molecular interactions at an unprecedented scale, reducing the time needed to develop new treatments. The financial sector is also exploring Quantum AI for risk analysis, portfolio optimization, and fraud detection, leveraging its ability to process vast datasets more efficiently than conventional AI models. Similarly, in logistics and supply chain management, Quantum AI could optimize routing, inventory distribution, and demand forecasting, enabling businesses to operate with greater efficiency and adaptability.

Although widespread quantum computing applications are still several years away, organizations can take proactive steps to prepare for this technological shift. Investing in quantum research and collaboration with academic institutions and quantum computing startups can provide valuable insights into potential applications. Businesses should also prioritize the development of quantum-safe encryption methods, as quantum computing will eventually pose a threat to existing cryptographic security protocols. By future-proofing their data security strategies, organizations can mitigate risks associated with the rise of quantum computing. Additionally, companies can begin exploring the practical applications of Quantum AI by conducting pilot projects and evaluating how quantum technologies may enhance their existing AI models.

These advancements are not just theoretical innovations but represent a fundamental shift that will shape the next industrial revolution. As illustrated in Figure 9-1, the integration of Quantum AI into manufacturing and other industries will redefine how complex computations are handled, leading to new efficiencies and groundbreaking solutions. By understanding and preparing for the impact of Quantum AI, businesses can position themselves at the forefront of this transformative wave, ensuring that they are ready to leverage its capabilities as the technology matures.

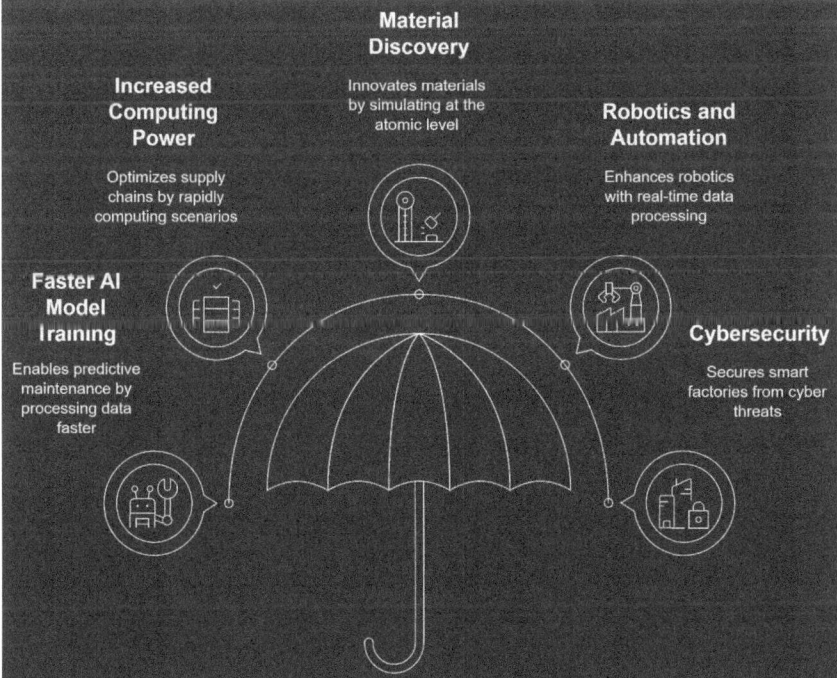

Figure 9-1: The future of AI: Quantum AI's potential impact on manufacturing

AI-Driven Transformation

Let's look at an example of an AI-driven transformation. A large multinational manufacturing company with production facilities in more than 15 countries faced increasing pressure to streamline its operations, reduce production costs, and maintain consistency in product quality across its global sites. The company produced a wide range of products, from automotive parts to consumer electronics, and operated complex supply chains to deliver raw materials and finished products across continents. Managing production and logistics efficiently in such a vast, interconnected operation was challenging, with issues ranging from fluctuating demand to supply chain disruptions.

To address these challenges and remain competitive in an increasingly digital world, the company turned to Gen AI to enhance its operations and optimize production processes. AI would play a pivotal role in improving everything from supply chain logistics to predictive maintenance while simultaneously driving operational efficiency and reducing waste across the company's global facilities.

The company realized that traditional methods of managing operations—such as human-driven decision-making and manual monitoring of production lines—were no longer sufficient for optimizing the global manufacturing process. To stay competitive, it deployed AI-powered models that were capable of analyzing massive datasets in real time, including historical production data, supplier performance metrics, weather forecasts, and even geopolitical risk indicators.

The AI models implemented across the company's sites were designed to be multimodal, allowing them to process and generate insights from different types of data simultaneously. For instance, the AI system could monitor production line sensors to track equipment performance while analyzing supply chain data to predict potential disruptions in material deliveries. With larger, more powerful AI models, the system provided real-time insights into plant performance, helping the company make data-driven decisions faster.

One significant advancement in AI's capabilities came through the use of few-shot and zero-shot learning models. These models allowed the AI systems to adapt to new production environments with minimal training data. For example, when the company opened a new facility in Southeast Asia, the AI model was able to optimize production efficiency by rapidly learning from the initial batches of production data without requiring months of detailed historical data. This flexibility was critical for a global company with diverse manufacturing environments and fluctuating demands.

Enhanced Personalization and Customer Engagement

Given the complexity of its global operations, the company used AI to improve its relationship with customers by offering personalized production timelines, customized order fulfillment, and tailored recommendations for new product offerings. The AI system analyzed customer preferences, regional market trends, and historical order data to provide clients with personalized production plans.

For example, if a customer in Europe placed a high-volume order for customized components, the AI system could predict the optimal production facility based on capacity, proximity, and lead times. It also offered insights into potential supply chain delays or fluctuations in material costs, providing the customer with real-time updates on their order. This level of personalization strengthened relationships with key clients, who valued the transparency and reliability of the AI-driven system.

To enhance customer satisfaction even further, the company introduced AI-powered virtual assistants that interacted directly with clients. These virtual assistants could answer complex queries about production schedules, material specifications, and customization options, providing instant responses based on real-time data. As a result, customer engagement increased and satisfaction scores rose, especially for large clients managing multiple product orders across different regions.

AI-Enhanced Decision-Making and Automation

A key benefit of Gen AI was its ability to support strategic decision-making at the executive level. The company's global leadership used AI-enhanced analytics to gain deep insights into manufacturing trends, equipment performance, and operational costs. AI's predictive capabilities allowed the company to anticipate market shifts, forecast demand, and optimize production schedules across its global network.

For example, the AI system helped the company automate inventory management across its facilities. The system analyzed demand patterns, supplier performance, and production capacity to maintain optimal inventory levels, reducing both excess stock and stockouts. This automation improved efficiency by 20%, as the AI system could dynamically adjust production schedules and material orders based on real-time market data, ensuring that production lines remained fully stocked without overcommitting resources.

In addition to inventory management, the company used AI-driven IPA to reduce manual intervention in routine tasks. The AI system handled complex workflows such as equipment monitoring, maintenance scheduling, and quality control checks. Predictive maintenance algorithms analyzed sensor data from production equipment to detect early signs of wear and tear, allowing the company to schedule maintenance before costly breakdowns occurred. This resulted in a 25% reduction in downtime across its facilities, improving overall equipment effectiveness (OEE) and reducing operational costs.

Ethical and Regulatory Challenges in AI Implementation

Operating in multiple countries, the company had to navigate a complex regulatory landscape while implementing AI across its sites. To ensure compliance with local and international regulations, the company adopted strict data governance policies, ensuring that all AI-driven decisions were made transparently and in accordance with regional data protection laws.

For example, in regions like Europe, where the GDPR governs data usage, the company implemented AI models with privacy protection features. These models were trained on anonymized data, ensuring that sensitive customer and employee information was not exposed during processing. The company also conducted regular audits of its AI systems to prevent bias in decision-making, particularly in areas like workforce allocation and hiring practices.

By adopting XAI technologies, the company was able to increase transparency around the decisions made by its AI systems. Executives and plant managers could review the rationale behind AI-driven recommendations, whether for operational improvements or strategic investments. This not only built trust in AI but also ensured that the company could meet regulatory requirements for accountability in high-stakes decision-making processes.

Through the adoption of advanced GenAI technologies, the global manufacturing company transformed its operations, improving efficiency, reducing costs, and enhancing customer engagement. By leveraging AI for product design, supply chain optimization, and personalized customer interactions, the company maintained a competitive edge in a rapidly evolving market. AI-driven insights and automation enabled the company to manage its global network of facilities with greater agility and precision, laying the groundwork for future advancements in Quantum AI and other emerging technologies.

Strategic Considerations for Leadership

As Gen AI continues to evolve, leaders must stay informed about emerging trends and proactively address the challenges associated with AI implementation. By embracing innovation and preparing for future advancements, organizations can leverage AI to gain a competitive edge and drive long-term success.

The following are a number of key recommendations for leadership to help in gaining leverage and competitive advantage:

- **Stay informed about AI developments:** AI is evolving rapidly, and staying up to date on the latest trends and technologies is critical. Engage with AI research, attend industry conferences, and collaborate with AI experts to understand how new innovations can benefit your organization.

- **Invest in AI talent and training:** As AI technologies evolve, so do the skills required to manage and implement them. Invest in training programs to upskill your workforce and ensure that they are equipped to work with next-generation AI tools.

- **Adopt an ethical AI framework:** Develop a comprehensive ethical AI framework that addresses issues like bias, transparency, and data privacy. This will help your organization navigate regulatory challenges and build trust with customers and stakeholders.

- **Explore emerging technologies:** Keep an eye on emerging technologies like Quantum AI and autonomous systems. Although these technologies may not be immediately accessible, early exploration and investment can position your organization as a leader in the next wave of AI innovation.

Leadership in AI Governance and Ethical Responsibility

In the previous chapters, we explored how organizations can strategically implement AI, ensuring that infrastructure, tools, and scaling methods align with business goals. Now, it's time to return to the structured playbook approach, focusing on one of the most critical aspects of AI deployment: governance and ethical responsibility.

For leaders in the manufacturing sector, AI presents unprecedented opportunities to drive innovation, optimize operations, and unlock new business value. From predictive maintenance and quality control to supply chain optimization and workforce automation, AI is transforming the industry. However, with great power comes great responsibility. The deployment of AI systems at scale brings ethical challenges, regulatory concerns, and risks that can impact an organization's reputation, legal standing, and long-term success.

Leadership plays a pivotal role in navigating these complexities. Leaders must not only ensure that AI technologies are adopted and implemented strategically but also lead the charge in establishing robust AI governance frameworks that prioritize ethical practices, transparency, and compliance with regulations. This chapter provides a structured framework for AI governance, tailored specifically for manufacturing leaders, outlining how to

- Establish AI governance frameworks to oversee AI adoption responsibly

- Develop ethical AI principles that ensure fairness, transparency, and accountability

- Implement compliance strategies to align AI initiatives with global regulatory requirements

- Build a culture of responsible AI use, ensuring that AI aligns with both business and societal values

By taking an active leadership role in AI governance, manufacturing organizations can foster innovation while maintaining public trust and regulatory compliance.

The Role of Leadership in AI Governance

Leadership is the cornerstone of effective AI governance. Without strong leadership, AI initiatives can easily drift off course, leading to unintended consequences such as data misuse, biased decision-making, or privacy violations. As manufacturing organizations increasingly rely on AI to automate processes, make decisions, and interact with suppliers and customers, it becomes essential for leaders to establish governance structures that oversee AI deployment, mitigate risks, and ensure compliance with ethical and legal standards. Figure 10-1 highlights the four keys to how leadership can impact and oversee AI governance.

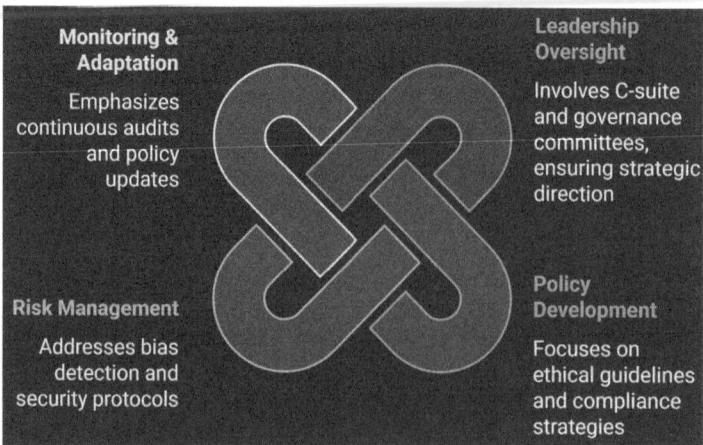

Figure 10-1: AI governance framework diagram

For manufacturing leaders, AI governance extends beyond technical considerations. While ensuring that AI systems function effectively, they must also integrate AI into broader strategic objectives while maintaining ethical integrity and regulatory compliance. This involves overseeing AI-driven quality control systems, ensuring that predictive maintenance models operate without unintended biases, and implementing workforce management tools that promote fairness and transparency. Effective AI governance requires a blend of vision, structure, and ethical leadership to balance innovation with trust, compliance, and operational integrity.

A key leadership responsibility is defining the organization's AI vision and governance structure. AI should not be seen merely as an automation tool but as a strategic driver of transformation. Leaders must articulate a clear vision for AI's role in the organization, ensuring that it aligns with business priorities and long-term objectives. This vision must be supported by governance structures that establish policies, roles, and oversight mechanisms to ensure AI is developed and deployed responsibly. In manufacturing, this could include setting quality control standards that ensure AI models used in production lines remain transparent, unbiased, and aligned with operational efficiency goals.

Beyond setting a strategic vision, leaders must establish a robust governance framework that spans the entire AI lifecycle—from data collection and model development to deployment and continuous monitoring. This involves forming governance committees responsible for evaluating AI applications across production, supply chain management, and predictive maintenance. For example, in manufacturing, auditing AI-driven defect detection models can help prevent unnecessary waste and production delays by minimizing false positives and ensuring accuracy.

Ethical leadership is equally critical in AI governance. Organizations must embed ethical considerations at every stage of AI development and implementation, ensuring that AI promotes fairness, transparency, and accountability. Leadership must actively communicate a commitment to responsible AI use, both internally to employees and externally to stakeholders, including suppliers, customers, and regulators. In manufacturing, this means preventing AI-powered workforce management systems from reinforcing biases in hiring, promotions, or performance evaluations. It also extends to sustainability, where AI-driven supply chain optimizations should support ethical sourcing and environmentally responsible manufacturing practices.

Accountability and oversight play a fundamental role in ensuring AI remains aligned with business and ethical standards. Leaders must implement mechanisms such as regular audits, compliance reviews, and bias-detection protocols to assess AI performance and mitigate risks. AI models should be continuously monitored to ensure they operate as intended and adapt to evolving regulatory expectations. In a manufacturing context, this could include auditing predictive maintenance systems to confirm that recommendations for equipment servicing are based on accurate, unbiased data, preventing unnecessary costs or unexpected failures.

Strong AI governance ensures that AI adoption fuels innovation while upholding ethical and regulatory commitments. By setting a clear vision, developing structured oversight frameworks, championing ethical AI leadership, and enforcing accountability, leaders can drive responsible AI deployment. As AI becomes more embedded in production, logistics, and workforce management, leadership will be instrumental in ensuring these systems enhance efficiency and competitiveness while remaining fair, compliant, and trustworthy.

The Need for Leadership-Driven AI Governance in Manufacturing

AI governance is not just a technical or operational concern—it is a strategic imperative that requires executive leadership. Without leadership-driven governance, AI systems can introduce significant risks, including perpetuating bias, exacerbating inequality, and violating privacy regulations. Additionally, a lack of governance can result in AI systems that are difficult to explain or audit, making it challenging to ensure fairness and accountability. Establishing a mature AI governance framework is essential for organizations that seek to implement AI responsibly while maintaining transparency in policy and decision-making. As illustrated in Figure 10-2, organizations must develop structured governance models that evolve alongside their AI capabilities, ensuring that they remain responsible and transparent in their AI strategies.

For manufacturing leaders, the stakes are particularly high. AI is increasingly embedded in critical functions such as supply chain management, quality control, and workforce automation. Without proper oversight, these AI-driven systems can lead to biased decision-making, regulatory violations, and reputational damage. Leadership-driven AI governance is essential for identifying risks early and implementing safeguards that prevent ethical breaches or unintended consequences. Ensuring that AI is aligned with corporate values, operational goals, and regulatory requirements is not just about compliance—it is about creating a foundation for ethical, sustainable, and competitive AI adoption.

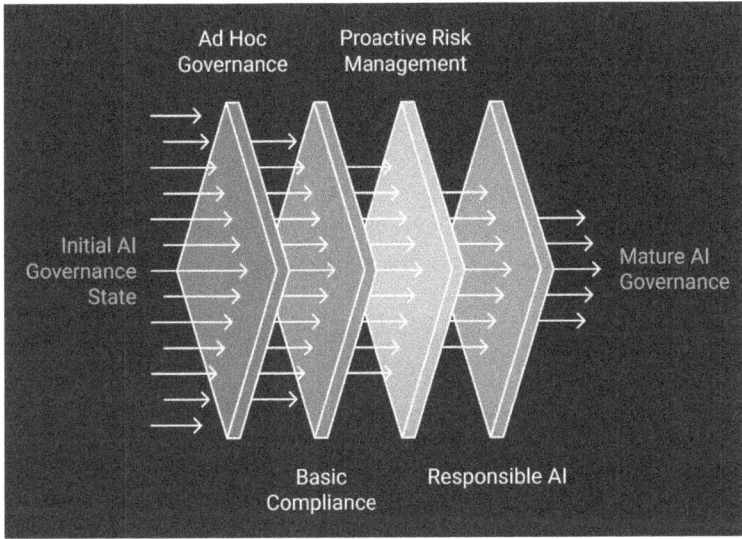

Figure 10-2: AI governance maturity progression

One of the most important roles of leadership in AI governance is mitigating risk. If AI technologies are left unchecked, they can expose organizations to legal penalties, ethical violations, and reputational harm. Strong AI governance frameworks help companies proactively manage these risks by implementing oversight mechanisms such as regular audits, algorithmic bias detection, and clear accountability structures. For instance, AI systems used in predictive maintenance in manufacturing must be routinely evaluated to ensure that they are not unintentionally prioritizing certain machines or work orders in ways that could lead to equipment failure or safety hazards. By setting governance standards early, leaders can prevent AI from introducing inefficiencies or safety concerns into critical operations.

Beyond risk management, leadership plays a critical role in building trust in AI-driven processes. Trust is a foundational element of AI adoption, both within the organization and in interactions with external stakeholders. AI systems that operate transparently and ethically are more likely to gain the trust of employees, customers, and regulatory bodies. Leadership fosters this trust by championing responsible AI practices and ensuring that AI-driven decisions are fair, explainable, and aligned with societal values. In manufacturing, this could mean ensuring that

AI-powered quality control systems operate with transparency, providing human operators with clear insights into how AI detects defects or irregularities. If employees can understand and challenge AI decisions when necessary, trust in AI systems will grow, leading to smoother adoption and better overall performance.

Leadership is also essential in ensuring the sustainability of AI initiatives. AI technologies evolve rapidly, and organizations must be prepared to adapt to these changes in a way that is both scalable and aligned with long-term goals. Leaders must ensure that AI governance frameworks are not rigid but flexible enough to accommodate new regulations, ethical standards, and technological advancements. In manufacturing, this might mean ensuring that AI-driven supply chain management systems can adapt to evolving environmental sustainability regulations, such as carbon footprint tracking or ethical sourcing requirements. A well-governed AI strategy should not only support current operational needs but also anticipate future industry shifts, keeping the organization competitive and resilient.

Effective AI governance in manufacturing requires leaders to take a proactive approach, integrating AI oversight into corporate strategy, risk management, and ethical decision-making. By prioritizing transparency, accountability, and adaptability, leadership-driven AI governance creates a foundation for responsible AI adoption, ensuring that AI technologies enhance operations rather than introduce unintended risks. In a rapidly evolving technological landscape, manufacturing leaders who take governance seriously will not only safeguard their organizations but also position themselves as industry leaders in ethical and sustainable AI innovation.

Leadership's Role in Ethical AI Practices

As AI becomes increasingly embedded in manufacturing, leadership plays a critical role in ensuring responsible implementation. Ethical AI governance is not just about regulatory compliance—it is about maintaining public trust, fostering fairness, and ensuring long-term sustainability. Leaders must proactively address risks related to bias, data privacy, and accountability, integrating AI systems that align with both business goals and ethical principles. By establishing clear governance frameworks, they can create an AI-driven manufacturing environment that prioritizes responsible innovation while mitigating unintended consequences.

A major challenge in AI governance is bias. Because AI models learn from historical data, they can inherit and reinforce existing inequalities, leading to unfair outcomes in hiring, workforce management, and quality control. For instance, AI-driven recruitment tools may inadvertently favor certain demographic groups, and AI-powered defect detection could produce skewed results if training data lacks diversity. To counteract these risks, leaders must implement robust bias-mitigation strategies, including diverse training datasets, regular audits, and explainable AI models that provide transparency into decision-making.

Data privacy is another critical concern, as AI systems rely on vast amounts of operational and employee data from IoT-enabled equipment, supply chains, and workforce analytics. Ensuring compliance with regulations like GDPR and CCPA is essential, but ethical AI governance goes beyond legal requirements. Leaders must establish strong data governance policies that define how information is collected, stored, and used, ensuring sensitive data is protected while fostering a culture of responsible data stewardship.

Transparency and accountability in AI decision-making are fundamental for building trust with employees and stakeholders. AI models must be explainable, ensuring that human operators can interpret their recommendations, particularly in quality control and predictive maintenance. Leaders should advocate for AI systems that provide clear, auditable reasoning for their outputs, avoiding black-box algorithms that reduce trust and limit oversight. Establishing governance structures—such as AI ethics committees and compliance officers—ensures that AI remains aligned with ethical and operational standards.

As AI adoption accelerates, manufacturing leaders must integrate fairness, security, transparency, and accountability into every stage of AI deployment. By championing responsible AI practices, they can build a sustainable AI-driven future where technology enhances business integrity, operational efficiency, and societal well-being.

Leadership Strategies for Ethical AI Governance in Manufacturing

Effective AI governance requires leadership to balance innovation with ethical responsibility and regulatory compliance. To ensure AI is implemented responsibly, leaders must establish clear policies, promote transparency, and embed ethical principles into daily operations.

This proactive approach minimizes risks related to bias, data privacy, and accountability while maximizing AI's benefits.

Addressing AI bias is a key priority, as models trained on incomplete or unbalanced datasets can reinforce systemic inequalities. Leaders must implement ongoing bias-detection frameworks that assess AI models throughout their lifecycle, ensuring fair outcomes in workforce management, hiring, and production decisions. Collaborating with domain experts and integrating diverse datasets enhances AI's ability to deliver equitable results.

Data privacy is another cornerstone of ethical AI governance. AI-driven manufacturing systems process vast amounts of data, from IoT-enabled machinery to workforce analytics. Leaders must ensure that AI applications comply with privacy regulations and are designed to protect sensitive data. Beyond compliance, organizations should foster a culture of ethical data use, ensuring that employees understand how AI systems process and analyze personal or operational data.

Transparency and explainability in AI decision-making strengthen trust and accountability. In areas like quality control and supply chain management, AI models must provide interpretable outputs so human operators can validate AI-generated decisions. Leaders should mandate the use of explainable AI and ensure that teams can challenge and refine AI-driven insights as needed.

Finally, accountability structures must be well-defined within the organization. Assigning clear oversight roles to AI governance committees, compliance officers, and operational managers ensures AI decisions are monitored and continuously refined. Employees should have pathways to escalate AI-related concerns, reinforcing a culture of accountability where ethical considerations are prioritized at every level of AI adoption.

By implementing these leadership strategies, manufacturers can integrate AI in ways that are both innovative and ethical. Ensuring fairness, safeguarding data privacy, and promoting transparency will not only mitigate risks but also build trust and long-term sustainability in AI-driven operations.

Building an Ethical AI Culture in Manufacturing

Beyond policies and governance structures, ethical AI leadership requires fostering a culture where responsible AI use is embedded in every aspect of operations. An ethical AI culture ensures that all employees—from data scientists to factory workers—understand the importance of fairness, transparency, and accountability in AI applications. This cultural shift begins with leadership, as executives and managers set the tone for AI adoption across the organization.

Promoting fairness and inclusion is a fundamental aspect of ethical AI leadership. AI models should be trained to serve all users equitably, which requires diverse datasets and proactive bias mitigation. In manufacturing, this is particularly relevant for workforce management systems, where AI may influence hiring, performance evaluations, and promotions. Leaders must ensure that these AI-driven processes are designed to promote inclusion rather than reinforce systemic biases.

Transparency in AI decision-making is equally essential. Employees and regulators must be able to understand how AI systems reach their conclusions, particularly in safety-critical industries. In manufacturing, AI-driven quality control systems must provide clear explanations for defect identification, allowing human operators to verify decisions and intervene when necessary. Leaders should advocate for AI models that offer interpretability, ensuring that human oversight remains a central component of AI-driven workflows.

Accountability is another critical pillar of an ethical AI culture. Leaders must ensure that AI teams, engineers, and decision-makers are responsible for the ethical implications of AI-driven outcomes. This involves establishing structured review processes, conducting regular audits, and addressing ethical concerns as they arise. In manufacturing, predictive maintenance AI systems must be monitored to ensure that their recommendations are accurate and do not create unintended inefficiencies or risks.

By integrating fairness, transparency, and accountability into the company's AI strategy, leadership can cultivate an ethical AI culture that fosters trust, innovation, and long-term success. Ethical AI practices must become a core part of business operations, ensuring that AI-driven processes enhance efficiency while aligning with human values and societal expectations. The responsibility of AI governance extends beyond technical teams—it requires active participation from leadership to drive awareness, enforce standards, and ensure AI serves both business objectives and broader ethical principles.

Case Study: Leadership in AI Governance at a Global Manufacturing Company

A global manufacturing company known for its high-volume production and stringent quality control standards faced significant challenges in ensuring that its AI-driven quality control systems operated fairly and transparently. As AI took on a larger role in automating defect detection and production monitoring, leadership recognized the need to establish a governance framework that would address concerns related to bias, explainability, and accountability. Without clear oversight, there was a risk that AI systems could introduce inconsistencies in product evaluations, potentially impacting supplier relationships, customer trust, and regulatory compliance.

To tackle these challenges, the company's leadership took a proactive approach by forming a cross-functional AI governance committee that included representatives from data science, compliance, operations, and quality assurance teams. This committee was tasked with overseeing AI deployments across the production process, ensuring that all AI-driven decisions aligned with the company's ethical standards and business objectives. One of their first initiatives was the implementation of bias-detection tools designed to regularly audit AI models for any disparities in how they assessed product quality. The company discovered that some AI models were more likely to flag defects in products made with certain raw materials, even though there was no meaningful difference in actual defect rates. This bias could have led to unfair treatment of suppliers and unnecessary material waste. By refining the AI algorithms and diversifying training datasets, leadership ensured that product quality assessments were conducted equitably, regardless of the origin or composition of materials used.

Another key focus for the leadership team was transparency in AI decision-making. They implemented a requirement for all AI quality control systems to generate clear, explainable outputs that could be easily understood by human workers. Previously, quality control managers were often presented with AI-generated decisions without context, making it difficult to challenge or verify results. With the new governance policies in place, AI models were modified to provide detailed explanations for each defect classification, including the specific patterns or anomalies that triggered a rejection. This explainability initiative not only helped build trust among quality control teams but also improved collaboration between AI systems and human inspectors, allowing for more effective oversight and reducing unnecessary product rejections.

In addition to internal governance improvements, the company extended its AI accountability practices to its supplier and customer relationships. Suppliers were given greater visibility into how AI systems evaluated raw materials, reducing disputes over quality assessments and fostering stronger partnerships. Customers, too, benefitted from the enhanced transparency, as the company was able to provide verifiable AI-driven quality assurance reports, demonstrating compliance with safety and performance standards. These efforts strengthened the company's reputation for fairness and reliability in the market.

The results of these leadership-driven AI governance initiatives were significant. Product defects were reduced by 20%, as AI models became more accurate and unbiased in their assessments. The company also saw a reduction in supplier disputes, as AI-driven quality control decisions were more consistent and explainable. Employee engagement improved, with greater confidence in AI's role in production processes, as workers were able to better understand and oversee AI-driven assessments.

This case study underscores the critical role of leadership in AI governance. By taking ownership of ethical AI implementation, the company not only enhanced its quality control processes but also built a culture of transparency, accountability, and trust. Its experience highlights the importance of cross-functional collaboration, bias mitigation, and explainability as key components of responsible AI governance in manufacturing. Through these efforts, the company positioned itself as a leader in ethical AI adoption, ensuring that AI remained a strategic asset rather than a potential liability.

Conclusion: AI Governance as a Competitive Advantage in Manufacturing

AI governance is not just about risk mitigation: it is a competitive advantage. Manufacturing organizations that embed governance into AI strategy from the start will

- Ensure that AI systems remain transparent, fair, and accountable
- Protect against regulatory fines and reputational risks
- Foster trust among customers, employees, and investors

As AI continues to shape the manufacturing industry, organizations that lead in governance will lead in innovation. AI governance is not optional—it is the foundation for responsible and sustainable AI-driven success.

Measuring AI Success: KPIs and Metrics for Evaluating Impact

As organizations increasingly integrate Gen AI into their operations, it becomes critical to measure the success of AI initiatives. Simply deploying AI is not enough; leadership needs to evaluate the return on investment (ROI), track performance improvements, and ensure that AI solutions align with overall business goals. This requires establishing clear metrics and key performance indicators (KPIs) that reflect the impact AI is having on the organization's processes, customer experiences, and bottom line.

In this chapter, we explore the most important KPIs and metrics for assessing the effectiveness of AI projects, how to align these metrics with business objectives, and how leaders can continuously refine AI systems based on performance data.

Defining Success: Aligning AI Metrics with Business Goals

The first step in measuring AI success is to define what success looks like for the organization. AI metrics should not only track the technical performance of the AI systems but also reflect the strategic objectives of the business. It is crucial that AI KPIs align with the broader business goals, whether those involve cost reduction, revenue growth, customer satisfaction, or operational efficiency. For example, if a company's goal is to improve customer satisfaction, the business objectives for AI should focus on aspects such as response times, resolution rates, and customer feedback.

When aligning AI metrics with business goals, it is essential to consider how AI fits into the organization's core objectives. AI initiatives should directly support business priorities, and the selected metrics should reflect this. Each department within the organization may have different AI-related goals based on its specific needs. For instance, the operations department may focus on metrics related to efficiency and process automation, and the marketing team may prioritize metrics related to personalization and customer engagement.

There is also a distinction between tangible and intangible outcomes of AI. Tangible benefits refer to measurable, quantifiable outcomes that AI delivers, such as increased efficiency, cost savings, and revenue growth. These are typically tracked using performance metrics and financial data, making them easier to evaluate. Examples include faster processing times, reduced labor costs, higher production output, and fewer errors in automated tasks.

Intangible benefits, on the other hand, are less quantifiable but equally valuable advantages that AI brings to an organization. These benefits impact creativity, strategic decision-making, and organizational culture but may not have immediate numerical representation. Examples include enhanced innovation, better employee engagement, improved decision-making, and stronger brand reputation due to AI-driven personalization or ethical AI use.

By recognizing both tangible and intangible outcomes, organizations can develop a holistic view of AI's value, ensuring that they account for both immediate financial gains and long-term competitive advantages. Some benefits, like improved operational speed or reduced manual workloads, are relatively easy to quantify and measure. However, other advantages, such as enhanced innovation and improved decision-making, may be more difficult to capture with standard metrics and could require more qualitative evaluation to gauge their impact on the business.

To ensure that AI metrics align with business goals, organizations should start by identifying clear business objectives. This requires collaboration with key stakeholders across departments to clarify the specific goals AI projects are intended to support. Whether the focus is on reducing operational costs, enhancing customer experience, or driving revenue growth, these objectives will guide the choice of relevant AI metrics. Once the objectives are defined, selecting the right KPIs is crucial. For example, if optimizing inventory management is a key goal, relevant KPIs may include inventory turnover rates or the frequency of stockouts.

Finally, since AI initiatives often affect multiple functions within the business, it's important to involve stakeholders from various departments in the process of selecting and refining KPIs. This cross-departmental collaboration ensures that each team has its own set of AI performance metrics, tailored to its specific role within the organization, while still contributing to the overall business objectives. By aligning AI metrics with strategic business goals and ensuring cross-functional collaboration, organizations can effectively measure the success of their AI initiatives. This framework can be seen in Figure 11-1. This chronological path will allow you to effectively set up meaningful and value-adding KPIs and metrics.

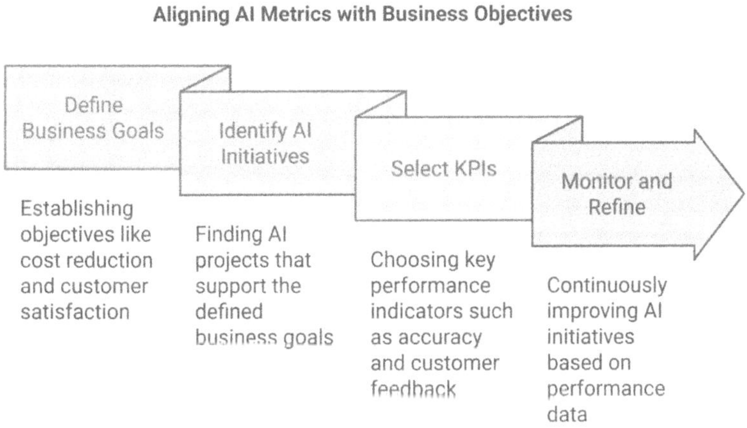

Aligning AI Metrics with Business Objectives

Define Business Goals	Identify AI Initiatives	Select KPIs	Monitor and Refine
Establishing objectives like cost reduction and customer satisfaction	Finding AI projects that support the defined business goals	Choosing key performance indicators such as accuracy and customer feedback	Continuously improving AI initiatives based on performance data

Figure 11-1: Framework for aligning AI metrics with business goals

Key Performance Indicators for Evaluating AI Success

Once business objectives are established, the next critical step is identifying the specific key performance indicators that will track the performance and impact of AI. These KPIs should be a mix of quantitative and qualitative metrics, reflecting various aspects of AI's effectiveness, such as efficiency, accuracy, scalability, and user experience. A well-rounded set of KPIs will provide a comprehensive picture of how AI is contributing to business goals, ensuring that both the tangible and intangible benefits are captured.

Quantitative KPIs are essential for measuring the direct, measurable outcomes of AI implementations. Quantitative KPIs are numerically measurable metrics that track the direct, objective impact of AI. These KPIs rely on hard data, statistics, and financial figures, making them easier to analyze and compare over time. Accuracy and precision are crucial metrics in this regard. Accuracy measures how closely the AI model's predictions align with real-world outcomes, which is especially important in applications like predictive maintenance. For instance, in a system designed to predict equipment failures, accuracy would indicate the proportion of correct maintenance predictions, helping to ensure that repairs are made before breakdowns occur. Precision, on the other hand, measures how often the AI system's predictions are correct when it identifies a specific condition—like predicting equipment failure in cases where it was necessary. Both of these metrics ensure that AI models are not only accurate but also reliable in their predictions.

Cost reduction is another key quantitative KPI, as AI is often implemented with the goal of reducing operational costs. Whether by automating tasks, optimizing resource allocation, or improving process efficiency, AI can lead to significant savings. Relevant KPIs may include cost per transaction, overall cost savings due to automation, and reductions in labor costs as tasks are streamlined through AI. Measuring these outcomes provides clear evidence of AI's contribution to making the organization more cost-efficient.

Process efficiency is also an area where AI's impact can be quantitatively measured. KPIs in this category track how AI improves the speed and effectiveness of different business processes. For example, a company may monitor the time-to-completion for AI-automated tasks, reductions in manual processing time, or improvements in supply chain throughput. These efficiency gains are often a primary reason that organizations invest in AI. Tracking these KPIs helps ensure that AI systems are delivering on their promises.

Revenue growth is a KPI that ties directly to AI's role in increasing the company's top line. In cases where AI is used in marketing, sales, or customer service, its ability to drive revenue becomes a critical measure of success. KPIs could include sales growth attributed to AI-driven initiatives, such as personalized marketing campaigns, increases in upsell or cross-sell opportunities facilitated by AI, and reductions in customer acquisition costs due to AI-enhanced customer targeting.

In addition to quantitative KPIs, qualitative KPIs offer valuable insights into the broader impact of AI on the organization. Qualitative KPIs, in contrast, focus on subjective, experience-based insights that are not always easily measured in numbers but still play a crucial role in evaluating AI's impact. One important qualitative metric is customer satisfaction. AI can greatly enhance customer experience, whether through chatbots that handle inquiries efficiently or recommendation systems that tailor shopping experiences. Customer satisfaction KPIs may include metrics like the net promoter score (NPS), customer feedback on AI interactions, and average response times for AI-driven customer support. These KPIs help the company gauge how AI is influencing customer loyalty and brand perception.

Employee engagement is another qualitative KPI that reflects AI's impact on the workforce. As AI takes over repetitive and time-consuming tasks, employees are freed up to focus on more strategic, value-added work. Monitoring employee engagement can provide insights into whether AI is improving job satisfaction or enhancing workplace culture. Relevant KPIs may include employee feedback on AI tools, the perceived ease of use of these systems, and the overall impact of AI on employees' daily workflows. These metrics help ensure that AI implementation leads to a more productive and satisfied workforce. In Table 11-1, you will see a brief example of some key metrics by business area for those leveraging different AI tools.

Table 11-1: Example Key AI Metrics by Business Function

BUSINESS FUNCTION	KEY AI METRIC	WHY IT MATTERS
Customer service	AI-driven response time (sec)	Tracks efficiency of AI virtual assistants
Operations	Process automation rate (%)	Measures AI's impact on reducing manual workloads
Marketing and sales	AI-driven conversion rate (%)	Evaluates AI's effectiveness in personalizing campaigns
Supply chain	Inventory prediction accuracy (%)	Measures AI's ability to optimize stock levels
Finance and compliance	Fraud detection success rate (%)	Determines AI's role in reducing financial risk

Innovation and creativity are harder to quantify but remain vital qualitative KPIs, especially for organizations aiming to use AI as a catalyst for growth. AI can enhance innovation by speeding up ideation, design processes, or problem-solving. For instance, AI can assist in generating new product ideas or rapidly testing different solutions in a fraction of the time it would take using traditional methods. KPIs in this area may include the number of new product ideas or innovations generated with AI assistance, or the reduced time it takes to bring AI-enhanced products to market.

By focusing on a combination of quantitative and qualitative KPIs, organizations can thoroughly assess AI's contribution to both operational and strategic goals, ensuring that the technology is not only improving performance metrics but also driving innovation, employee satisfaction, and customer engagement.

Measuring ROI for AI Initiatives

Return on investment is a key metric for evaluating the financial success of AI initiatives. Calculating ROI for AI projects involves a comprehensive comparison between the financial benefits generated by the AI system and the costs incurred during its deployment and ongoing operation. Understanding ROI is critical for organizations looking to determine the value AI adds to their business and whether the investment is justified.

One of the first factors to consider when calculating AI ROI is the overall cost of deployment. This includes the initial investment required to purchase or develop AI technologies, set up the necessary infrastructure, and hire or train the talent needed to manage and maintain the system. It's important to account for ongoing costs as well, such as system maintenance, continuous data management, and periodic retraining of AI models. These costs can add up over time and should be factored into the total investment to ensure an accurate calculation of ROI.

To begin calculating ROI for an AI project, organizations should first estimate the total costs involved. This includes both the upfront expenses related to technology acquisition, infrastructure, and staffing, as well as the ongoing operational costs. These estimates provide a clear picture of the total investment required. Once the costs are defined, the next step is to quantify the financial benefits that AI is expected to deliver. For example, if AI automates a labor-intensive process, the cost savings from reduced manual work should be calculated. If AI helps increase sales through better customer targeting, the resulting revenue growth should be assigned a monetary value.

Quantifying the benefits of AI is equally important and involves assigning financial value to the improvements brought about by AI implementation. These benefits often include direct cost savings through automation, revenue growth due to improved customer engagement, and operational efficiency gains from AI optimizing processes. Some benefits, such as more informed decision-making and increased innovation, may be more challenging to quantify but can still play a crucial role in determining the overall success of an AI project. Although these intangible benefits may not always have a direct financial value, they should be considered when assessing the broader impact of AI on the organization.

Another critical aspect of measuring AI ROI is the time horizon. Unlike other investments, AI initiatives may take time to deliver measurable results. Some AI systems, particularly those focused on long-term process improvements, may not generate immediate returns. It's essential to establish a realistic timeline for evaluating the ROI of an AI project, understanding that some benefits may become apparent only after the system has been fully integrated and optimized. This could mean setting milestones for quarterly or annual evaluations, depending on the nature of the AI initiative.

Finally, organizations should compare the costs and benefits over the established time horizon. By analyzing how AI-generated benefits stack up against the costs over time, whether on a quarterly, annual, or project-based timeline, businesses can determine whether the AI initiative is delivering a positive ROI. This comparison provides valuable insights into whether the project is financially viable in the short term or if it will provide more substantial returns over a longer period. A clear understanding of ROI helps businesses make informed decisions about whether to scale AI solutions, invest in additional AI technologies, or adjust their AI strategies to maximize returns.

In summary, calculating AI ROI requires a detailed assessment of both the costs and the financial benefits of AI implementation. By estimating total investment, quantifying the financial returns, and evaluating the project over a realistic time horizon, organizations can determine whether their AI initiatives are delivering the expected value and driving long-term success.

Ongoing Monitoring and Continuous Improvement

Measuring AI success is not a one-time activity but an ongoing process. AI systems evolve as business conditions change, new data is introduced, and models need to be retrained. Because AI performance can fluctuate over time, continuous monitoring is essential to ensure that AI systems remain aligned with business objectives and continue to deliver value. Without regular oversight, AI models may become less effective, and their outputs may no longer provide reliable insights.

One of the key considerations in monitoring AI success is the phenomenon of *model drift*. Over time, AI models can become less accurate as the underlying patterns in the data they were trained on change. For example, a customer behavior model that was trained on past purchasing data may become outdated as market trends shift or as new customer segments emerge. By implementing continuous monitoring, organizations can quickly detect and address model drift, ensuring that the AI systems remain reliable and continue to produce accurate results. When model drift is detected, follow the corrective actions outlined in Table 11-2.

Table 11-2: AI Model Drift Indicators and Countermeasures

DRIFT INDICATOR	IMPACT	COUNTERMEASURE
Drop in prediction accuracy	AI outputs become unreliable.	Retrain the model with updated data.
Increase in false positives	AI misclassifies.	Adjust decision thresholds.
Decline in user engagement	AI recommendations lose relevance.	Improve personalization algorithms.

Retraining AI models is another critical aspect of ongoing AI success. As businesses gather new data, it's important to update AI models to reflect this latest information. The frequency of retraining will depend on the nature of the AI system and the dynamic nature of the data it processes. For instance, AI models that handle rapidly changing data—such as sales trends or customer interactions—may require more frequent retraining than those used in relatively stable environments. Regularly updating models ensures that AI systems remain relevant and continue to provide actionable insights that align with current business conditions.

In addition to technical monitoring, establishing feedback loops from employees and customers who interact with AI systems is crucial. Users often have direct experience with the AI's outputs, whether through internal operations or customer-facing services like chatbots or recommendation engines. Gathering feedback from these users allows organizations to identify potential gaps or issues in the AI's performance. For example, if customers consistently report dissatisfaction with the responses they receive from an AI-powered virtual assistant, this feedback can help refine the system to better meet their needs. By incorporating user feedback, organizations can ensure that their AI systems are not only performing well technically but also delivering value from a user experience perspective.

To ensure ongoing success, organizations should implement real-time monitoring of their AI systems. Real-time monitoring tools enable businesses to track key performance metrics continuously, allowing for immediate identification of any issues that arise. Setting up alerts for critical KPIs can help organizations respond quickly if there is a drop in accuracy, efficiency, or other performance measures. For instance, if an AI-driven supply chain optimization tool suddenly shows lower-than-expected results, real-time alerts can notify management to investigate and address the issue before it impacts operations.

Establishing feedback loops is another essential practice for maintaining AI success over time. This involves regularly collecting feedback from users and analyzing their experiences with the AI system. Feedback from both employees and customers offers insights into how the AI is performing in real-world scenarios and can reveal areas where improvements or adjustments are needed. For example, an AI tool used for automating customer service responses may need to be adjusted if customers frequently report misunderstandings or irrelevant suggestions. Regularly reviewing this feedback ensures that AI models stay in tune with user expectations and business needs.

Finally, organizations should schedule reviews of AI performance at set intervals, such as quarterly or annually. During these reviews, businesses can assess whether their AI models need retraining, whether new data sources should be incorporated, or whether KPIs need to be refined to reflect changes in business priorities. These periodic assessments ensure that AI systems remain aligned with long-term business goals and continue to deliver meaningful results. For instance, a quarterly review may reveal that an AI system used for marketing campaign optimization needs to incorporate new social media trends to stay relevant and effective.

In summary, ongoing monitoring of AI success involves real-time tracking, regular feedback loops, and scheduled performance reviews. By staying proactive and continuously refining AI models, businesses can ensure that their AI systems remain accurate, relevant, and aligned with evolving business objectives, providing long-term value across the organization.

Qualitative Feedback and Human Oversight

Although quantitative KPIs offer clear, measurable insights into the performance of AI systems, qualitative feedback is just as important for gaining a holistic understanding of how AI impacts an organization. Human oversight is crucial to ensuring that AI models are used effectively and responsibly. Regular evaluations from employees and customers who interact with AI systems help ensure that these tools complement human decision-making and align with the organization's values. Qualitative feedback also provides context to the numbers, offering insights that metrics alone may miss.

One key area for qualitative feedback is user experience. It's essential to assess how both employees and customers engage with AI systems. Are the tools intuitive and easy to navigate? For employees, AI systems should ideally improve productivity and streamline workflows, whereas for customers, they should enhance the overall experience, whether through faster service or more personalized interactions. Collecting feedback on user experience helps identify any friction points where the AI system may be falling short or causing frustration. For example, if a customer-facing chatbot is difficult to interact with, customers may express dissatisfaction even if the AI is technically performing well. These insights are crucial for making necessary adjustments that lead to better user engagement.

Another important area for feedback is decision support. AI is often deployed to enhance human decision-making by providing data-driven insights, but it's essential to evaluate whether the AI is truly helping employees make better decisions. Feedback from employees can reveal whether they trust the AI's recommendations and whether those insights are actionable. For instance, in a retail setting, AI may suggest inventory changes based on sales patterns, but employees may feel the suggestions don't fully account for local factors. Understanding how AI influences decision-making, and whether employees feel empowered by it, ensures that AI tools are effectively supporting the workforce rather than creating confusion or distrust.

Ethical considerations are also critical when gathering qualitative feedback. AI systems can raise concerns about data privacy, potential biases, and the transparency of their decision-making processes. Employees and customers can provide valuable feedback on whether they feel the AI systems are being used ethically and in line with the organization's values. For instance, if an AI system for recruitment is perceived as biased or unfair in its decision-making, this feedback can prompt the organization to reexamine and adjust the model to ensure that it is fair and compliant with ethical standards. Monitoring the ethical implications of AI use is essential to maintaining trust and upholding the organization's integrity.

To effectively incorporate qualitative feedback into AI evaluations, organizations should begin by conducting surveys and interviews. Surveys with open-ended questions can help gather insights from employees and customers about how they perceive the AI systems they interact with. Interviews offer a deeper exploration of specific issues or concerns, allowing organizations to understand the human perspective on AI's strengths and weaknesses. For example, employees using an AI tool for sales forecasting may share that although the tool is accurate, it lacks flexibility in adapting to market fluctuations, suggesting areas for improvement.

Additionally, organizations can establish human oversight committees made up of key AI users. These committees serve as a formal mechanism for collecting and reviewing feedback on AI systems. The committee members can meet regularly to discuss potential issues, recommend system improvements, and ensure that AI tools are being used responsibly. This oversight is especially important for maintaining accountability and ensuring that AI systems continue to serve the organization's goals and ethical standards. For example, in a healthcare setting, an oversight committee may regularly review the effectiveness and fairness of an AI tool used for diagnostics, ensuring that it consistently improves patient outcomes without introducing bias.

In summary, although quantitative KPIs are essential for measuring AI performance, qualitative feedback is equally important for ensuring that AI systems are used responsibly and effectively. By gathering feedback on user experience, decision-making support, and ethical concerns, organizations can ensure that their AI systems align with human values and enhance rather than hinder business operations.

Key Takeaways for Leadership

Measuring the success of AI initiatives is essential for ensuring that AI delivers tangible value to the organization. By selecting the right KPIs, tracking both quantitative and qualitative metrics, and continuously refining AI systems, leaders can ensure that AI contributes to strategic business objectives and generates a strong return on investment.

Key recommendations for leadership include the following:

- **Align AI KPIs with business objectives** to ensure that AI initiatives support core goals.

- **Track both quantitative and qualitative metrics** to gain a comprehensive view of AI's impact on the organization.

- **Continuously monitor AI performance** and adjust models as needed to maintain accuracy and relevance.

- **Incorporate human oversight** to ensure that AI systems are used responsibly and ethically.

Overcoming Common Challenges in AI Adoption

Although the potential benefits of Gen AI are immense, the journey to successful AI adoption is often riddled with challenges. Organizations may face a variety of obstacles, ranging from technical hurdles to organizational resistance. Understanding these challenges is the first step toward overcoming them.

In this chapter, we will explore some of the most common barriers to AI adoption and provide practical strategies for addressing them, ensuring that AI initiatives are successfully implemented and sustained. Figure 12-1 gives a brief overview of some of the common challenges AI adoption faces and what we will cover in this chapter.

Data Quality and Availability

One of the most critical components of any AI initiative is data. AI systems rely on large volumes of high-quality, well-structured data to deliver accurate and meaningful results. However, many organizations struggle with data that is incomplete, outdated, inconsistent, or siloed across different departments. Poor data quality can undermine the effectiveness of AI models and limit the potential benefits of AI.

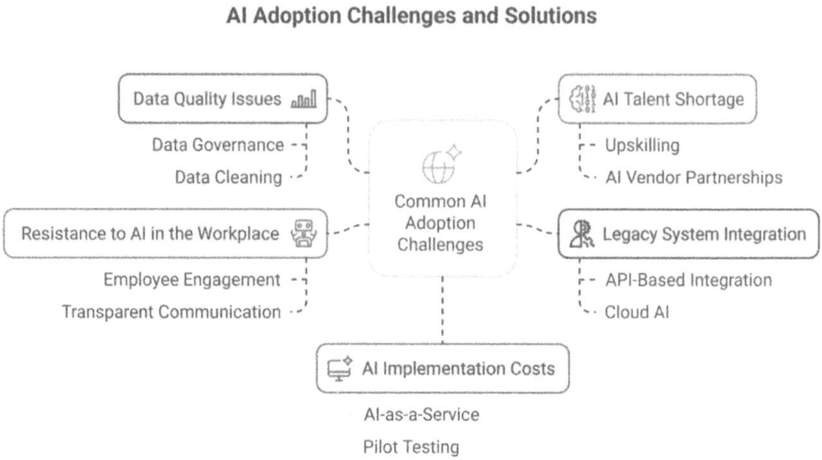

AI Adoption Challenges and Solutions

```
Data Quality Issues                    AI Talent Shortage
   Data Governance                        Upskilling
   Data Cleaning                          AI Vendor Partnerships
                         Common AI
                         Adoption
Resistance to AI in the Workplace   Challenges    Legacy System Integration
   Employee Engagement                     API-Based Integration
   Transparent Communication               Cloud AI

              AI Implementation Costs
                 AI-as-a-Service
                  Pilot Testing
```

Figure 12-1: Common challenges in AI adoption and their solutions

A major challenge in AI adoption is the existence of data silos, where information is stored in different systems or departments, making it difficult to access and integrate. When data is scattered across multiple platforms, AI systems may not have a complete view of the information they need, leading to inaccurate predictions and inefficiencies. Additionally, poor data quality presents a significant hurdle, as inconsistent or inaccurate data can compromise AI model performance, resulting in biased or incorrect outputs. Without clean and structured data, AI cannot generate reliable insights, making data integrity a crucial aspect of any AI strategy. Another important consideration is data privacy concerns, as organizations must comply with strict regulations like GDPR and CCPA when handling sensitive information. These regulations may restrict access to personal data, limiting the amount of training data available for AI models and requiring organizations to adopt stringent data protection measures.

To overcome these challenges, organizations must invest in a robust data management infrastructure. Implementing data lakes or data warehouses allows companies to consolidate and standardize data across departments, ensuring that AI systems have access to high-quality, structured information. This approach streamlines data integration and enhances the reliability of AI-driven insights. Additionally, data cleaning and preprocessing should be a priority before deploying AI models. This includes eliminating duplicate entries, filling in missing data, and ensuring that data formats are consistent across different systems. By refining data inputs, organizations can improve the accuracy and performance of their AI applications.

Ensuring data governance and compliance is also essential for maintaining data integrity and security. Organizations should implement robust data governance policies that define how data is collected, stored, and used while ensuring compliance with privacy regulations. Establishing clear guidelines for handling personal data and implementing tracking systems to monitor data usage can help organizations align with legal standards and build trust with stakeholders. By prioritizing data management, quality control, and regulatory compliance, organizations can create a strong foundation for AI success, unlocking its full potential while minimizing risks.

Talent Gaps and Lack of Expertise

A significant challenge in AI adoption is the shortage of specialized talent. Developing and deploying AI systems requires expertise in areas such as data science, machine learning, and AI ethics. Many organizations, particularly small and medium-sized enterprises, struggle to find and retain the skilled professionals needed to effectively implement AI initiatives. Without the right expertise, organizations risk underutilizing AI technology or encountering difficulties in managing AI-driven processes.

One of the primary obstacles is the shortage of AI talent. The demand for AI professionals far exceeds the available supply, making it difficult for companies to attract and hire skilled data scientists, AI engineers, and machine learning specialists. Compounding this issue is the lack of in-house expertise, as many organizations do not have the technical knowledge required to design, implement, and maintain AI solutions. Without internal AI proficiency, businesses may struggle with model deployment, data management, and algorithm optimization. Another challenge is the high cost of AI talent. Due to the competitive nature of the field, hiring experienced AI professionals can be expensive, which can be a major barrier for smaller companies that lack the budget to build a dedicated AI team.

To overcome these challenges, organizations can upskill their existing workforce. Investing in employee training programs that focus on data analysis, machine learning, and AI ethics can help bridge the knowledge gap and enable current staff to take on AI-related roles. Providing access to certifications, workshops, and online courses allows employees to build AI expertise without the company having to recruit externally. Another effective approach is to partner with AI vendors or consulting firms that offer turnkey AI solutions and ongoing support. These partnerships allow organizations to leverage the expertise of specialized AI providers without the need to hire full-time AI professionals.

Additionally, businesses can take advantage of AI development platforms such as Google Cloud AI, Microsoft Azure, or AWS AI. These platforms provide prebuilt AI models and tools that can be customized to meet specific business needs, reducing the requirement for deep in-house AI expertise. Many of these platforms also offer support services that assist with implementation, allowing companies to deploy AI solutions without requiring extensive internal technical knowledge.

By focusing on upskilling employees, collaborating with AI vendors, and leveraging AI platforms, organizations can navigate the talent gap and successfully integrate AI into their operations. These strategies help companies remain competitive in an AI-driven world while minimizing the challenges associated with hiring and retaining specialized AI talent.

Integration with Legacy Systems

Integrating AI into existing legacy systems presents a significant technical challenge for many organizations. Many companies continue to rely on outdated infrastructure that may not be compatible with modern AI solutions. Ensuring that AI systems can function seamlessly alongside legacy systems is critical for unlocking the full value of AI adoption. However, legacy infrastructure can create bottlenecks, slow processing speeds, and limit the ability of AI to provide real-time insights.

One of the main challenges organizations face is outdated infrastructure. Many legacy systems were not designed to handle the computational requirements of AI models, particularly those that rely on real-time data processing or large-scale machine learning operations. These limitations can prevent businesses from fully leveraging AI's predictive capabilities or automation potential. Additionally, integration complexity poses another major hurdle. AI models often need to interact with multiple databases, software applications, and platforms, creating compatibility issues. Ensuring seamless data exchange between AI and existing systems requires technical expertise and a structured approach to integration. Finally, the cost of upgrading infrastructure can be a significant barrier. Replacing or modernizing legacy systems can be expensive and disruptive to business operations, making organizations hesitant to embark on large-scale system overhauls.

To overcome these challenges, organizations should adopt a phased approach to integration. Instead of overhauling all legacy systems at once, businesses can start by identifying high-impact areas where AI can deliver immediate value. For example, AI can be implemented in customer service automation or predictive maintenance before being expanded into more complex workflows. This step-by-step approach minimizes disruption while allowing organizations to assess AI's impact before making further investments.

Another effective strategy is to use API-based integration. Application programming interfaces (APIs) allow AI models to communicate with existing software and databases without requiring a complete system replacement. By leveraging APIs, businesses can enable AI to extract insights from legacy systems, automate workflows, and improve decision-making without disrupting current operations. This approach ensures that AI models can function within the organization's existing technology stack while gradually enhancing system capabilities.

For organizations looking to bypass the limitations of legacy infrastructure, cloud-based AI solutions offer a viable alternative. Cloud computing platforms, such as AWS, Microsoft Azure, and Google Cloud AI, provide scalable computing power and data storage without requiring major infrastructure upgrades. By migrating AI workloads to the cloud, businesses can leverage real-time processing capabilities, access advanced AI tools, and integrate AI-driven insights into their operations without investing in costly hardware upgrades. Additionally, cloud solutions allow for easier updates and maintenance, ensuring that AI applications remain cutting-edge without the need for frequent system overhauls.

By phasing AI integration, utilizing APIs, and leveraging cloud based AI solutions, organizations can successfully modernize their operations while minimizing disruptions. These strategies enable businesses to harness the power of AI without incurring the high costs and complexities associated with upgrading legacy systems, allowing them to remain competitive in an increasingly AI-driven world.

Resistance to Change and Organizational Culture

The adoption of AI often requires organizations to redefine workflows, decision-making processes, and company culture. Although AI has the potential to enhance efficiency and drive innovation, employees may resist its implementation due to fears of job displacement or a lack of understanding about how AI works. To successfully integrate AI, businesses must foster a culture that views AI as a tool for augmentation rather than replacement and encourages employees to engage with new technologies rather than fear them.

One of the most significant challenges in AI adoption is fear of job loss. Employees may worry that AI will replace their roles, making them hesitant or resistant to its implementation. This fear can lead to low morale, reluctance to engage with AI tools, and even pushback against AI-driven initiatives. Another common challenge is lack of understanding. Many employees do not fully grasp how AI functions, how it impacts their work, or the benefits it can bring to their daily tasks. Without proper education, employees may view AI as an unnecessary disruption rather than an opportunity for growth. Additionally, some organizations struggle with cultural resistance to change. If a company has traditionally been slow to adopt new technologies, employees and leadership alike may be skeptical of AI and resist integrating it into established workflows.

To overcome these challenges, leadership must take a proactive approach to addressing AI-related concerns. The first step is to communicate the benefits of AI clearly and transparently. Employees need to understand how AI will enhance their roles rather than replace them. Leaders should emphasize how AI automates mundane, repetitive tasks, freeing employees to focus on more strategic, creative, and high-value activities. When employees recognize that AI is a supportive tool rather than a threat, they are more likely to adopt it with an open mind.

Another key strategy is to actively involve employees in AI initiatives. Instead of rolling out AI systems without employee input, organizations should engage workers in the decision-making process. This can include training programs, workshops, and pilot projects that allow employees to interact with AI tools before they are fully implemented. When employees see firsthand how AI can improve their work, they are more likely to embrace its adoption. Additionally, offering training and development programs ensures that employees gain the necessary skills to work alongside AI, making them feel empowered rather than displaced.

Fostering a culture of innovation is also essential for overcoming AI resistance. Organizations should encourage employees to experiment with AI tools, providing safe spaces for trial and error. Recognizing and celebrating small AI successes can help build confidence and demonstrate the value of AI in the workplace. Leadership should serve as role models by demonstrating a commitment to AI adoption, showing employees that AI is not just an initiative but a strategic priority for the company.

Overcoming resistance to AI is a challenge that all organizations will face at different stages of their AI journey. Regardless of where a company stands in its AI adoption process, it is critical to implement structured steps to manage change effectively. Figure 12-2 outlines five key steps that organizations can follow to streamline the process of overcoming AI resistance. There is no one-size-fits-all solution, but these steps provide a strategic framework to help employees transition into an AI-enabled future with confidence and enthusiasm.

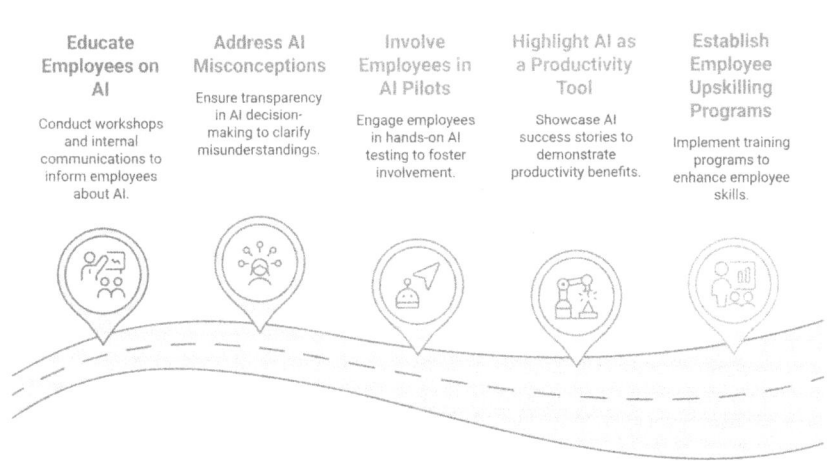

Figure 12-2: Five steps to overcoming resistance to AI in the workplace

Managing the Costs of AI Implementation

AI implementation can be a significant financial investment, requiring resources for technology infrastructure, data management, skilled talent, and ongoing system maintenance. Many organizations, particularly small and medium-sized businesses, may hesitate to adopt AI due to concerns about the high upfront costs and uncertainty surrounding return on investment (ROI). To successfully integrate AI without straining financial resources, businesses must develop strategic cost-management approaches that allow them to maximize AI's value while keeping expenses under control.

One of the primary financial challenges of AI adoption is the high initial costs associated with deployment. Implementing AI requires investments in data infrastructure, cloud computing, software development, and

AI-specific talent. Organizations often struggle to justify these expenses, particularly when the ROI is not immediately visible. Another major concern is the uncertainty surrounding ROI. Businesses may be reluctant to invest in AI if they are unsure when they will see measurable benefits. Unlike traditional technology investments, AI requires ongoing model training, performance optimization, and integration efforts, which may delay financial returns. Additionally, ongoing costs for AI maintenance and upgrades must be accounted for. AI models are not static; they require continuous refinement, retraining, and compliance with evolving industry regulations. These factors contribute to the long-term cost of AI adoption and must be carefully managed.

To mitigate these financial challenges, organizations should start with small-scale pilot projects before committing to full-scale AI implementation. By launching targeted AI pilots in high-value areas, businesses can test AI's effectiveness in real-world scenarios without making significant upfront investments. A successful pilot project can demonstrate AI's impact on efficiency, automation, and decision-making, helping to justify further investments to key stakeholders.

Another effective cost-management strategy is to leverage cloud-based AI solutions instead of investing in expensive on-premise infrastructure. Cloud platforms, such as Google Cloud AI, Microsoft Azure AI, and AWS AI, provide scalability and flexibility, allowing businesses to pay only for the resources they use. Cloud-based AI also eliminates the need for heavy IT infrastructure investments, reducing capital expenditures while providing access to cutting-edge AI tools and capabilities.

To ensure that AI investments generate long-term value, organizations must continuously measure ROI using key performance indicators (KPIs). By tracking cost savings, revenue growth, and efficiency improvements, businesses can quantify the financial benefits of AI implementation and make data-driven decisions about future investments. Regular ROI evaluations allow companies to adjust AI strategies, optimize spending, and allocate resources more effectively.

Managing AI implementation costs requires a balance between strategic investment and cost control. By starting with pilot projects, utilizing cloud-based AI solutions, and continuously measuring AI's financial impact, organizations can minimize financial risks while maximizing AI-driven benefits. AI does not have to be a prohibitive expense—with the right financial planning, it can become a sustainable and value-generating asset for the organization.

Examples of Overcoming Common Challenges

A good way to understand challenges and solutions when it comes to AI adoption is to consider real-life examples. In this section, we'll look at two different examples of challenges that businesses had and the solutions they adopted.

Example 1: Data Quality and Availability

Let's first look at an example of a challenge regarding data quality and availability. A large retail company wanted to implement AI-driven demand forecasting to optimize inventory and improve customer satisfaction. However, the organization quickly encountered a major roadblock: its data was spread across various departments, with sales, marketing, and supply chain each using different systems to store and manage their data. The lack of data integration made it difficult to build accurate AI models. Additionally, the company's customer data was incomplete and inconsistent, further complicating the process.

> **Challenge**   The retail company faced significant issues with data silos and poor data quality. Sales data was stored in an outdated system, and marketing campaigns were tracked separately, with limited coordination. Moreover, much of the customer data was either incomplete or outdated, resulting in unreliable inputs for the AI system. This led to inaccurate demand forecasts, increasing the risk of stockouts and excess inventory.

> **Solution**   The company decided to invest in a data management infrastructure to consolidate its scattered data sources. It implemented a centralized data warehouse that integrated sales, customer, and supply chain data, making it easier for AI systems to access clean, consistent, up-to-date information. Additionally, the company used data cleaning and preprocessing techniques, which involved eliminating duplicate entries, filling in missing values, and standardizing formats across all departments.

> By implementing data governance policies, the company also ensured that it was compliant with data privacy regulations such as GDPR, protecting sensitive customer information while still making it available for AI training. This approach allowed the AI models to generate more accurate forecasts, ultimately reducing inventory costs and improving customer satisfaction.

Example 2: Talent Gaps and Lack of Expertise

As a second example, let's look at a mid-sized financial services firm. This firm recognized the potential of using AI to automate risk assessments and enhance fraud detection. However, the company struggled to find qualified AI talent. With limited in-house expertise in machine learning and data science, the firm found it difficult to design, implement, and maintain sophisticated AI systems. Additionally, the high salaries demanded by AI specialists were outside the firm's budget, making it challenging to fill the talent gap.

Challenge   The firm faced a shortage of AI talent and lacked the internal expertise necessary to build and scale AI systems. Hiring data scientists and machine learning engineers proved to be both difficult and expensive. Without the right team in place, the company's AI projects stalled, resulting in slow progress on its digital transformation goals.

Solution   To address the talent gap, the firm decided to upskill its existing employees. It provided training programs in data science and machine learning, focusing on key roles within the company that could benefit from AI capabilities, such as business analysts and IT staff.

By investing in internal development, the company not only improved its AI expertise but also fostered employee engagement by offering career growth opportunities. Additionally, the firm partnered with an external AI vendor to gain access to ready-made AI solutions and ongoing technical support. The vendor provided the firm with a customized fraud detection model and offered training for the team on how to manage and monitor the system.

By leveraging these AI development platforms and vendor expertise, the company was able to implement AI-driven fraud detection successfully without the need for an expensive, fully in-house AI team. This approach reduced risk while enabling the company to move forward with its AI initiatives.

These examples illustrate how organizations can tackle common AI adoption challenges, such as overcoming data silos and talent shortages, by implementing strategic solutions that involve investing in data infrastructure, upskilling employees, and partnering with external experts.

Key Takeaways for Leadership

Adopting AI can be transformative, but it requires careful planning and a strategic approach to overcome common challenges. By addressing issues related to data quality, talent, system integration, cultural resistance, and cost management, leaders can ensure that AI initiatives are successful and deliver long-term value.

Key recommendations for overcoming AI adoption challenges include the following:

- Invest in data management to ensure that AI models are built on high-quality, accessible data.

- Upskill employees and partner with external experts to address AI talent gaps.

- Use APIs and cloud-based solutions to integrate AI with legacy systems without costly infrastructure overhauls.

- Promote a culture of innovation by involving employees in AI initiatives and addressing concerns about job displacement.

- Start with pilot projects to reduce risk and demonstrate AI's value before scaling across the organization.

Implementing AI Governance and Ensuring Accountability

As organizations become increasingly reliant on AI systems to drive business outcomes, leadership plays a pivotal role in ensuring that these systems are governed effectively and used responsibly. *AI governance* refers to the frameworks, policies, and practices that ensure AI systems operate in a transparent, accountable, and ethical manner. It goes beyond deploying AI tools; it involves establishing oversight mechanisms that regulate how AI models are developed, trained, and monitored to prevent unintended consequences. Strong governance ensures that AI aligns with organizational objectives while adhering to ethical standards and regulatory requirements.

With the growing power and reach of AI, leaders must actively embed governance practices into their organization's culture and operations, ensuring AI innovation can flourish without compromising ethical values or stakeholder trust. AI-driven decisions can have profound and far-reaching consequences, particularly in high-stakes fields such as finance, healthcare, and law enforcement. In finance, biased AI models could lead to unfair lending practices, denying loans or credit to certain groups based on flawed data. In healthcare, AI-driven diagnostic tools must be rigorously tested to avoid misdiagnoses that could negatively impact patient outcomes. In law enforcement, AI-powered surveillance and predictive policing systems raise ethical concerns about privacy violations, racial profiling, and wrongful convictions.

The implications of AI governance extend beyond these industries. Unregulated AI systems can exacerbate biases, create security vulnerabilities, and reduce public trust in technology. Poorly governed AI could lead to misinformation, manipulation, or even safety risks in autonomous systems such as self-driving cars. Without clear governance structures, organizations face increased risks of regulatory penalties, reputational damage, and ethical lapses that could hinder AI adoption.

This chapter explores the key components of AI governance, focusing on leadership's responsibility in fostering an ethical AI culture, ensuring regulatory compliance, and driving accountability across all levels of the organization. Leaders must establish clear guidelines for data ethics, bias mitigation, and transparency to ensure that AI is deployed responsibly and contributes positively to both business and society.

Establishing a Comprehensive Governance Framework

A well-rounded AI governance framework is essential for ensuring that AI systems operate responsibly. Leadership must spearhead the creation of this framework, which should cover the entire lifecycle of AI—from development to deployment, monitoring, and scaling. A key leadership responsibility is to ensure that the framework integrates ethical guidelines, compliance protocols, and accountability structures to safeguard the company and its stakeholders. The key elements of an AI governance framework are outlined next. These elements can be seen in Figure 13-1.

It is imperative to say that without a robust framework, you will not be able to execute your AI strategy successfully. I also want to state that you need to remember to not stifle innovation; I cannot stress this enough.

Key elements of an AI governance framework:

- **Ethical AI guidelines:** Leaders must set clear ethical principles that govern AI initiatives. These guidelines should address fairness, bias, and transparency, ensuring that AI systems do not perpetuate discrimination or harm certain groups of people. Ethical guidelines provide a moral compass for AI projects and help protect the company from reputational damage.

- **Data privacy and security:** Leadership must ensure that AI systems comply with rigorous data privacy and security measures, particularly when handling sensitive information. Compliance with data protection laws like GDPR and CCPA is paramount. By establishing clear data governance policies, leaders can protect both the organization and individuals from data breaches or misuse.

- **Continuous monitoring and compliance:** Leaders should put in place mechanisms for identifying and eliminating bias in AI algorithms. These mechanisms must include regular audits of AI models to detect and correct any biases that may lead to unjust or unequal outcomes, especially in areas like hiring, lending, and healthcare.

- **Accountability and oversight:** AI systems must be transparent and explainable, with clear mechanisms in place to hold individuals or teams accountable for the outcomes they produce. Leadership should ensure that AI decision-making processes can be easily audited and that explanations are available to both internal stakeholders and external regulators.

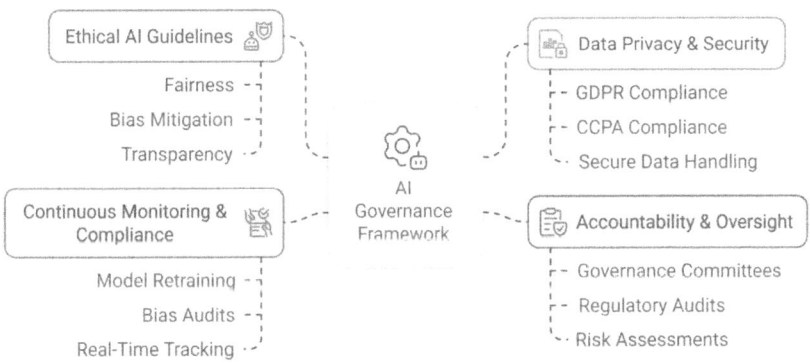

Figure 13-1: AI governance framework: the pillars of responsible AI

Leadership's Role in AI Governance

AI governance starts with leadership, as executives must set the tone for ethical AI use and lead by example. Leaders are responsible for creating a governance culture that prioritizes fairness, accountability, and compliance across the entire organization. To ensure that AI systems are aligned with these values, leadership must take a proactive approach, fostering an ethical culture while providing the necessary resources for proper governance implementation. As mentioned, leadership responsibilities do not fall on one person or one group: as shown in Table 13-1, AI governance is a shared responsibility.

Table 13-1: Leadership Responsibilities in AI Governance

LEADERSHIP ROLE	KEY AI GOVERNANCE RESPONSIBILITIES	WHY IT MATTERS
CEO and C-suite	Set AI ethical principles, oversee AI strategy alignment	Ensures that AI reflects company values and objectives
Chief AI/data officers	Ensure AI compliance, establish AI risk policies	Mitigates AI-related legal and ethical risks
AI governance committees	Monitor AI systems, review ethical concerns	Provides oversight for fairness, transparency, and bias mitigation
IT and engineering teams	Implement security protocols, data governance	Ensures that AI is built responsibly and securely
HR and compliance teams	Conduct AI training, manage AI bias audits	Prevents AI discrimination in hiring, promotions, etc.

Leadership responsibilities in AI governance include a number of things including

- Championing ethical AI
- Establishing governance committees
- Ensuring regulatory compliance
- Fostering a transparent culture

Leaders must advocate for ethical AI practices both internally and externally. This involves promoting AI fairness, transparency, and data privacy in meetings, decisions, and strategies. Leadership's vocal commitment to ethical AI signals to employees that responsible AI use is a non-negotiable priority.

Additionally, leadership should establish cross-functional AI governance committees responsible for overseeing AI initiatives. These committees, consisting of legal, IT, operations, and compliance experts, ensure that AI deployments are ethically sound and compliant with industry regulations. Leaders play a critical role in empowering these committees and ensuring that their recommendations are integrated into business decisions.

Ensuring regulatory compliance is also a key responsibility for leaders. They must maintain oversight of AI systems to ensure compliance with the latest legal and regulatory standards. Regular audits and reviews should be conducted to prevent any legal or ethical violations. Leaders must stay informed about evolving AI regulations and industry-specific compliance needs, and they must provide the resources necessary to ensure that AI projects are in line with these requirements.

Finally, leaders must foster a transparent culture. Transparency is a cornerstone of responsible AI use, and leaders must model this value throughout the organization. AI decisions should be explainable to all stakeholders—employees, customers, and regulators. By fostering a culture of transparency, leadership builds trust and ensures that AI systems enhance, rather than erode, stakeholder confidence.

Steps for Leaders to Implement AI Governance

To effectively govern AI, leadership must take concrete steps to integrate governance frameworks into their business practices. These steps ensure that governance is treated not as an afterthought but as a core part of the organization's strategy and operational model. These steps are more of a workflow; we talk about continuous improvement with AI, and this is no exception. Figure 13-2 shows this lifecycle.

The following explains the lifecycle steps that are presented in Figure 13-2:

1. **Develop clear AI policies and standards:** Leaders must collaborate with legal, technical, and operational teams to draft clear AI governance policies. These policies should detail how AI systems will be developed, monitored, and scaled, and they must include protocols for data privacy, transparency, and accountability. Creating a well-defined governance framework is the foundation for responsible AI use.

2. **Embed governance in the corporate strategy:** AI governance cannot exist in isolation; it must be integrated into the broader corporate strategy. Leaders must ensure that governance frameworks align with business goals and that AI initiatives adhere to these guidelines at every stage. This strategic alignment reinforces the

importance of responsible AI within the company's mission and values.

3. **Create a culture of accountability:** Leaders need to foster a culture in which employees understand their accountability in AI projects. By promoting discussions on ethical AI use and creating safe spaces for raising concerns, leadership can ensure that accountability is shared across the organization.

4. **Commit to continuous improvement:** The field of AI is evolving rapidly, and governance frameworks must evolve alongside it. Leaders must stay abreast of advancements in AI technology, ethics, and regulation to ensure that governance structures remain relevant and effective. Regularly updating governance policies demonstrates a commitment to long-term AI sustainability.

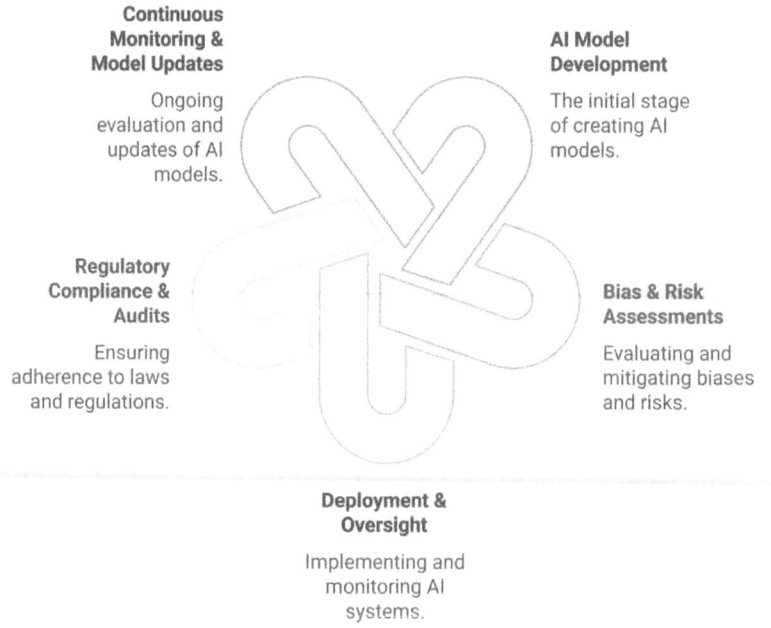

Figure 13-2: AI governance lifecycle: from development to compliance

Ethical Leadership and Regulatory Compliance

Compliance with legal regulations and ethical standards is crucial as organizations scale AI initiatives. Leadership must prioritize responsible AI use by enforcing these regulations and ensuring that AI systems align with societal values. As AI technologies advance, the potential for ethical breaches grows, and leaders must act decisively to safeguard against issues such as data misuse, bias, and privacy violations.

There are a number of key regulatory and ethical considerations leaders should take into account. These include

- **Data privacy laws:** With data at the core of AI systems, complying with data privacy laws like GDPR and CCPA is non-negotiable. Leadership must ensure that data is handled securely and ethically, with robust protections in place to avoid breaches and maintain consumer trust.

- **Bias and fairness:** AI systems can inadvertently introduce or perpetuate bias, especially when trained on unbalanced datasets. Leaders must ensure that AI systems are rigorously tested for bias and that strategies are in place to promote fairness in decision-making processes.

- **Explainability and accountability:** In industries where AI is used to make high-stakes decisions, such as healthcare, finance, and criminal justice, AI systems must be explainable. Leaders are responsible for ensuring that AI models can be understood, scrutinized, and validated, particularly when decisions affect individuals' lives or livelihoods.

Leadership's Role in Balancing Innovation and Governance

Leadership faces the challenge of balancing AI innovation with responsible governance. Although ethical and regulatory compliance is essential, it should not stifle innovation or limit the organization's ability to compete in the AI-driven marketplace. Effective leadership finds ways to encourage creative AI solutions while maintaining the guardrails necessary to prevent harm. Strategies for balancing innovation and governance include

- **Encouraging responsible innovation:** Leaders should promote innovation within ethical boundaries, encouraging teams to experiment with AI while considering its societal impact. By framing governance as a framework for responsible innovation rather than a limitation, leaders can create an environment where AI projects flourish within ethical guidelines.

- **Supporting cross-functional collaboration:** Leadership must facilitate collaboration between technical teams, such as data scientists, and business leaders, ensuring that AI initiatives meet both strategic and ethical objectives. This collaboration helps ensure that AI systems are scalable, compliant, and aligned with the company's goals.

- **Emphasizing responsible risk-taking:** Innovation often requires taking calculated risks. Leaders should empower teams to take responsible risks, knowing they operate within ethical boundaries. This involves creating space for experimentation while ensuring adherence to governance protocols.

Case Study: Leadership in AI Governance for a Global Manufacturing Firm

A global manufacturing company recently implemented AI-driven predictive maintenance systems to optimize machinery performance. Although the technology promised cost savings and increased efficiency, leadership recognized the need for strong governance to ensure responsible use. The company's executives established an AI governance committee composed of leaders from IT, compliance, operations, and HR. This committee developed ethical guidelines to govern AI use, addressing issues like bias in predictive models and data security concerns.

To further promote governance, leadership invested in AI literacy programs that trained managers and workers on how to interact with AI systems, interpret AI-driven recommendations and report any anomalies. By actively involving leadership in the governance process and ensuring transparency at all levels, the company successfully scaled its AI initiatives without compromising ethical standards. The predictive maintenance system resulted in a 20% reduction in equipment downtime while maintaining regulatory compliance and adhering to ethical AI principles.

Leadership is the cornerstone of effective AI governance. By setting the tone for ethical AI use, establishing comprehensive governance frameworks, and balancing innovation with responsibility, leaders can ensure that AI systems deliver value without causing harm. As organizations continue to scale AI initiatives, leadership's role in fostering accountability, compliance, and transparency will be critical in safeguarding the organization's future in an AI-driven world.

Key Takeaways for Leadership

AI governance is a leadership responsibility that ensures AI systems operate transparently, ethically, and in compliance with regulations. A strong governance framework mitigates risks such as bias, security breaches, and regulatory violations while enabling responsible AI innovation.

- **Establish a governance framework:** Leaders must implement structures that guide AI development, deployment, and scaling. Governance should include ethical AI guidelines, data privacy measures, and bias mitigation strategies.

- **Ensure transparency and accountability:** AI systems must be explainable and auditable. Leadership should enforce clear decision-making processes to ensure AI outputs can be validated by internal teams, regulators, and stakeholders.

- **Maintain legal and ethical compliance:** With evolving regulations like GDPR and CCPA, leaders must proactively ensure compliance and prevent discriminatory AI practices that could result in legal and reputational risks.

- **Foster a culture of AI accountability:** Governance is a shared responsibility. Leaders must define clear roles across AI teams, compliance, HR, and IT to ensure ethical AI adoption throughout the organization.

- **Balance innovation with responsible risk-taking:** Governance should not stifle AI progress. Leaders must promote safe experimentation and AI-driven solutions while maintaining ethical and regulatory boundaries.

- **Invest in AI literacy and training:** Employees must understand AI's capabilities, risks, and governance policies. Leadership should implement AI education programs to ensure responsible AI use across the organization.

- **Continuously monitor and improve AI systems:** AI governance is ongoing. Leaders must establish monitoring mechanisms to track AI performance, bias, and compliance, ensuring that AI models remain reliable and up to date.

- **Lead by example:** Executives must demonstrate ethical AI leadership by integrating governance into AI strategy, reinforcing transparency, and advocating for responsible AI adoption.

AI governance is a strategic necessity, not just a compliance requirement. Organizations that invest in governance today will reduce risk, enhance trust, and drive long-term AI success while maintaining ethical AI adoption at scale.

Creating an AI-First Culture: Fostering Innovation and Adaptability

AI adoption is more than just a technological shift—it requires a cultural transformation. Whereas earlier chapters focused on AI strategy, governance, and workforce development, this chapter delves into the organizational mindset and cultural shifts required to embed AI deeply into daily operations. Unlike previous discussions that covered structural changes and leadership responsibilities, this chapter focuses on how to cultivate an AI-first culture that fosters innovation, experimentation, and adaptability.

An AI-first culture is not about adopting AI in isolated projects but about creating an environment where AI is a fundamental part of decision-making, collaboration, and problem-solving across all levels of the organization. It requires breaking down silos, encouraging employees to experiment with AI tools, and fostering cross-functional collaboration that integrates AI into every aspect of business operations. This chapter outlines how leaders can drive cultural transformation by reshaping mindsets, empowering employees, and establishing an innovation-first approach.

This chapter differs from earlier chapters in the following ways:

- **Not just AI adoption, but cultural transformation:** Earlier chapters explored AI implementation, governance, and leadership's role in oversight. This chapter focuses on the mindset shifts required to make AI an integral part of an organization's identity.

- **Encouraging experimentation and risk-taking:** Rather than just ensuring compliance and governance, this chapter explores how to create an environment where employees feel empowered to explore AI-driven solutions without fear of failure.

- **Cross-functional collaboration:** Although workforce development was covered previously, this chapter focuses on breaking down organizational silos and encouraging AI literacy across all departments to drive innovation at scale.

- **Embedding AI in everyday decision-making:** The goal is not just to introduce AI tools but to make AI a natural and default part of workflows, strategies, and problem-solving across teams.

Building Blocks of an AI-First Culture

To succeed in the AI-driven era, organizations must foster an innovation-ready culture that integrates AI across teams and functions. Figure 14-1 highlights the core building blocks necessary to create an AI-first organization.

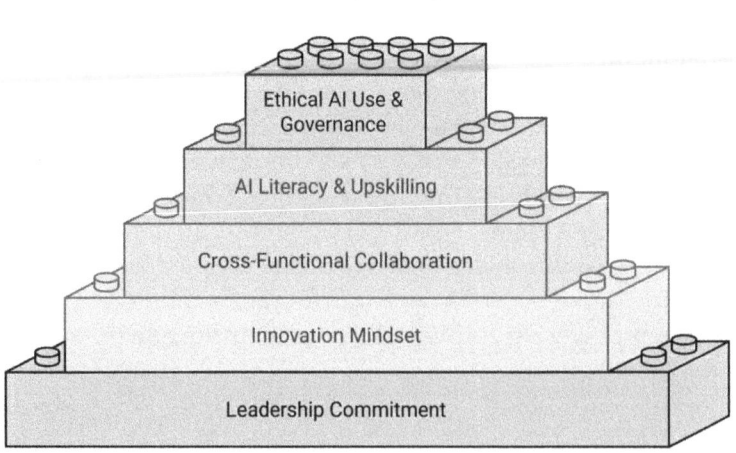

Figure 14-1: The AI-first culture framework

This chapter will explore specific leadership strategies to drive cultural change, encourage AI experimentation, and ensure that AI becomes a foundational element of daily business operations. By focusing on adaptability, learning, and collaboration, leaders can shape a future-ready workforce that thrives in an AI-powered world.

The Foundation of an AI-First Culture

At the core of an AI-first culture is the understanding that AI is more than just a tool for addressing specific problems—it is a catalyst for continuous innovation that can enhance every aspect of an organization. This culture emphasizes the integration of AI into decision-making processes, encourages employees to embrace new technologies, and fosters a mindset of ongoing experimentation and learning. By making AI an integral part of the company's DNA, organizations can unlock its full potential to drive innovation and growth.

A key element of an AI-first culture is leadership commitment. Leaders play a crucial role in setting the tone for how AI is perceived and adopted throughout the organization. Their active support for AI initiatives and their use of AI in strategic decision-making signal to employees that AI is a priority. When leaders model an AI-first mindset, it sets an example for the rest of the organization, reinforcing the importance of embracing AI as a strategic asset rather than just a technical tool.

Innovation at every level is another critical aspect of fostering an AI-first culture. An organization that truly embraces AI encourages innovation from all corners, not just from its leadership or specialized teams. Whether top executives or frontline employees, everyone should feel empowered to explore how AI can enhance their work and open up new possibilities. This democratization of AI fosters a culture where employees across departments feel comfortable experimenting with AI technologies, leading to more creative and impactful solutions.

Collaboration across functions is essential for successful AI adoption, as AI projects often require input from multiple teams, such as IT, operations, and business units. Breaking down silos between departments ensures that AI initiatives are well-rounded and address both technical and business needs. Cross-functional collaboration brings together different perspectives, making AI projects more effective and aligned with the organization's overall goals.

To build the foundation for an AI-first culture, leaders must first articulate a clear AI vision. This vision should explain how AI will drive innovation and enhance business outcomes while also aligning with broader organizational goals. By consistently communicating this vision across all levels of the company, leaders can ensure that employees understand the strategic role AI will play in the organization's future.

In addition to articulating a vision, leaders must demonstrate their commitment to AI by incorporating it into their own decision-making processes. Senior executives should actively participate in AI-driven initiatives, showing that AI is not just a buzzword but a core component of the company's long-term strategy. This visible commitment helps build trust and buy-in from employees, who will be more likely to embrace AI if they see leadership leading by example.

An AI-first culture encourages AI exploration at every level of the organization. Leaders should create opportunities for employees to experiment with AI through workshops, hackathons, or pilot projects. These initiatives not only help employees become more comfortable with AI tools and technologies but also spark creativity and new ideas for applying AI in different areas of the business. By promoting experimentation and learning, organizations can continuously evolve and improve their AI capabilities.

Building an AI-first culture requires leadership commitment, fostering innovation at all levels, and encouraging cross-functional collaboration. By articulating a clear AI vision, demonstrating leadership's dedication to AI, and promoting exploration and experimentation, organizations can fully integrate AI into their operations and unlock its potential as a driver of ongoing innovation.

Encouraging Experimentation and Risk-Taking

One of the hallmarks of an AI-first culture is a willingness to experiment and take calculated risks. AI's potential lies in its ability to generate new insights, automate processes, and solve complex problems, but it requires an environment that encourages trial and error. In many traditional business cultures, employees may be hesitant to try new approaches due to fear of failure or resistance to change. Leaders must create a culture where innovation is celebrated and experimentation is rewarded. This, for many companies, is an evolution; it's a process, and although it won't happen overnight, it is imperative that it is addressed and made a priority. Moving to an AI-first culture, as shown in Figure 14-2, is not just a novel concept: it is transformative, it is empowering, it is a condition of moving your business forward, but it is also met with the following key challenges:

- **Fear of failure:** Employees may be reluctant to engage in AI experimentation if they fear negative consequences for failure. This can stifle innovation and slow down AI adoption.

- **Risk aversion:** In organizations with a strong risk-averse culture, introducing AI-driven initiatives may face resistance from teams unwilling to deviate from established practices.

- **Lack of resources for experimentation:** Experimenting with AI requires access to data, tools, and technical expertise. Organizations must provide the necessary resources for employees to experiment and innovate with AI.

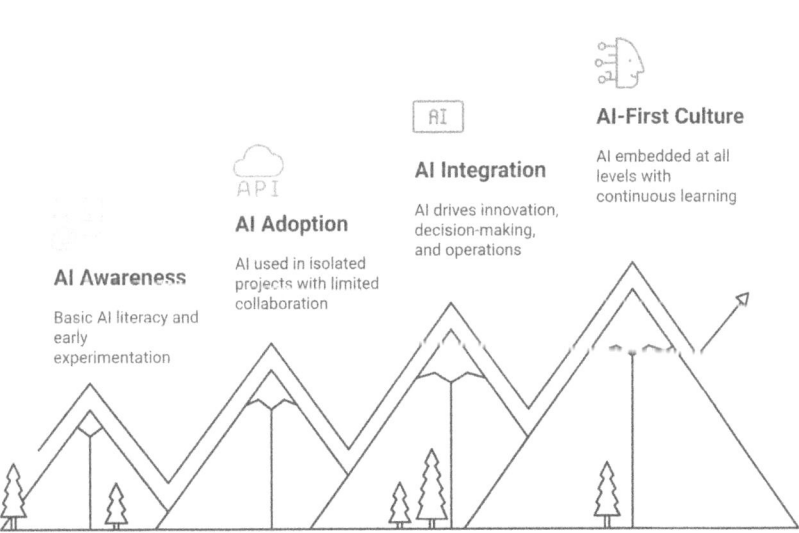

Evolution to an AI-First Culture

AI Awareness
Basic AI literacy and early experimentation

AI Adoption
AI used in isolated projects with limited collaboration

AI Integration
AI drives innovation, decision-making, and operations

AI-First Culture
AI embedded at all levels with continuous learning

Figure 14-2: The AI culture maturity model

Organizations that successfully foster experimentation establish structures and incentives that make AI-driven innovation part of everyday work. The following strategies help create a culture where AI experimentation is not only accepted but actively encouraged:

- **Create "safe spaces" for innovation:** Establish environments where employees can experiment with AI without fear of failure. This can involve innovation labs, hackathons, or pilot projects where teams can test new ideas and iterate on solutions.

- **Reward experimentation:** Encourage risk-taking by recognizing and rewarding employees who embrace AI-driven experimentation. Rewards don't have to be financial—public acknowledgment, career advancement opportunities, and the chance to lead future AI projects can be powerful motivators.

- **Provide resources for AI experimentation:** Ensure that teams have access to the necessary AI tools, data, and training to experiment effectively. This may include offering AI software licenses, datasets for testing, or partnerships with AI vendors for technical support.

Fostering a Culture of Experimentation in a Healthcare Organization Amid Regulatory Concerns

Let's look at an example. A large healthcare organization aimed to leverage AI to improve patient outcomes and streamline administrative tasks. However, the organization's culture was highly risk-averse due to the complexity of healthcare regulations and strict compliance requirements. Employees were hesitant to experiment with AI solutions because they feared that any misstep could lead to regulatory violations or legal issues. Leadership recognized that this cautious mindset was slowing innovation and AI adoption.

Key Challenge

In the past, healthcare professionals and administrators were reluctant to propose AI-driven initiatives out of concern for regulatory compliance. Fear of violating HIPAA or other healthcare regulations made employees hesitant to experiment with AI despite the potential benefits it could bring to patient care and operational efficiency. Additionally, there was a lack of resources, such as access to AI tools and compliance-focused technical expertise, which limited experimentation.

Strategy for Encouraging Experimentation

To address these concerns, the leadership team created "compliance-friendly innovation labs" within the organization. These labs served as safe spaces where employees could test AI applications in a controlled, regulation-compliant environment. For example, clinical teams could explore AI-driven diagnostic tools while working closely with the organization's legal and compliance experts to ensure that all experiments adhered to healthcare regulations. By involving compliance officers in these labs from the outset, the organization allowed teams to experiment with AI without the fear of inadvertently violating regulations.

Leadership also established a reward system to promote experimentation and risk-taking in a compliant manner. Employees who led AI-driven projects that adhered to regulatory standards were publicly recognized at staff meetings, and successful pilots were highlighted in internal newsletters. Healthcare professionals who embraced AI-driven experimentation were also offered career development opportunities, such as leadership roles in future AI projects or participation in high-profile AI healthcare initiatives. This recognition helped shift the culture, encouraging employees to see AI as an innovation tool rather than a regulatory risk.

Additionally, the organization ensured that employees had the necessary resources to experiment with AI effectively and within compliance boundaries. They provided access to AI tools tailored for healthcare settings, such as AI-powered diagnostic and administrative software that complied with HIPAA standards. The organization also partnered with AI vendors that specialized in healthcare, offering technical support and training that focused on regulatory-compliant AI implementation. Regular workshops and hackathons were held, where teams could collaborate on AI projects while receiving guidance from legal and compliance experts to ensure that all experiments met the required healthcare standards.

Outcome

Within a year, the organization saw a significant increase in AI experimentation, with projects carefully designed to comply with healthcare regulations. For example, one team successfully piloted an AI-driven patient triage system that reduced emergency room wait times by 20%. Additionally, an AI tool used to streamline billing and insurance claim processing led to a 15% reduction in administrative costs. By fostering a culture of compliant experimentation, the organization accelerated AI adoption while ensuring that all initiatives adhered to regulatory requirements, ultimately enhancing both patient care and operational efficiency.

Cross-Functional Collaboration and Continuous Learning

AI initiatives often require collaboration across various functions, including data science, IT, marketing, and operations. Building an AI-first culture necessitates breaking down traditional silos and fostering cross-functional teamwork, which allows organizations to leverage a diverse range of expertise. Cross-functional collaboration is key to creating more effective AI solutions, as it brings together different perspectives that help align AI projects with both technical and business objectives. However, achieving this level of collaboration can be challenging. In many organizations, teams work in isolation, making it difficult to share data, insights, or AI-driven innovations across departments. This lack of integration slows progress and limits the potential of AI initiatives. Moreover, poor communication between technical and nontechnical teams can hinder AI projects. Business units may struggle to grasp the full potential of AI, and data science teams may lack the necessary insight into specific business needs.

To address these challenges, organizations should create **cross-functional AI teams** that bring together employees from various departments, including data scientists, IT professionals, and business leaders. These teams can collaborate on AI projects, ensuring that the technical and business aspects of AI are aligned from the start. Regular communication is essential for keeping everyone on the same page. Establishing frequent check-ins or meetings to discuss AI initiatives, share progress, and address challenges promotes transparency and accountability. Collaboration tools that track projects and allow teams to communicate easily across functions can also help bridge the gap between teams.

Another strategy to improve collaboration is developing a **shared language** for AI. Nontechnical employees should have a basic understanding of AI concepts, while data scientists need to understand the specific business problems they are addressing. This shared language fosters more productive interactions and ensures that both technical and business teams can collaborate effectively.

A cornerstone of an **AI-first culture** is the commitment to continuous learning and skill development. As AI technologies rapidly evolve, organizations must ensure that their employees are constantly learning and adapting. Upskilling employees in AI literacy is essential for staying competitive in an AI-driven world. Many organizations face a **skills gap**, where employees may lack the technical know-how to work with AI systems, which can stall AI adoption. Additionally, the fast pace of AI advancements can make it challenging for organizations to keep their teams up to date. Continuous learning opportunities are necessary for employees to stay relevant and fully leverage emerging AI tools and techniques. Moreover, although some roles require deep technical expertise, others may need only a basic understanding of AI. Balancing AI literacy across various roles is crucial for ensuring that everyone in the organization can contribute to AI initiatives at the appropriate level.

To foster this culture of **continuous learning**, organizations should invest in AI training programs that are tailored to employees' specific roles. These programs should cover both technical skills, such as machine learning and data analysis, and broader AI literacy, helping employees understand AI's impact on business strategy. Encouraging lifelong learning is another critical component. Providing access to online courses, workshops, and certifications in AI-related fields helps employees keep their skills current. Incentivizing continuous learning through career advancement opportunities or internal recognition can further motivate employees to pursue additional training. Additionally, organizations can promote **mentorship** opportunities, where employees with AI expertise can mentor their colleagues. This approach not only spreads AI knowledge across the organization but also promotes collaboration and strengthens team bonds, creating a more cohesive and AI-literate workforce.

By fostering cross-functional collaboration and a commitment to continuous learning, organizations can build a strong AI-first culture that drives innovation and ensures that AI initiatives are both successful and scalable.

Measuring Cultural Impact and Adapting to Change

Fostering an AI-first culture is not a one-time effort; it requires continuous evaluation and adaptation as the organization evolves and AI adoption deepens. Leaders must regularly assess how well this culture is supporting both employees and business outcomes, ensuring that the organization remains flexible, innovative, and aligned with its AI goals. This process involves tracking key cultural metrics, gathering employee feedback, and making strategic adjustments to keep AI initiatives relevant and effective.

A primary indicator of a successful AI-first culture is **employee engagement**. It's essential for leaders to gauge how comfortable employees feel using AI tools and how actively they are participating in AI-driven initiatives. Employees who are engaged are more likely to embrace AI as a tool that enhances their work, rather than something that disrupts it. Monitoring engagement can also reveal how effectively the organization has integrated AI into its daily operations. Similarly, **adaptability and innovation** are important metrics for assessing AI culture. Organizations need to track how adaptable teams are to changes brought about by AI and how frequently AI-driven innovations are being proposed and tested. These factors indicate how well the organization is using AI to drive creativity and improvements across departments.

Feedback loops play a crucial role in measuring the effectiveness of an AI-first culture. Continuous feedback from employees helps leaders understand how AI tools and initiatives are impacting day-to-day operations. Employees should feel empowered to share their experiences—both positive and negative—regarding AI integration. This feedback provides valuable insights into where the organization's AI strategy is succeeding and where it may need adjustments. Leadership must be receptive to this input and ready to take action based on the feedback received.

To effectively measure and adapt an AI-first culture, organizations should begin by regularly **surveying employees**. Employee engagement surveys and AI adoption assessments provide concrete data on how employees perceive AI in their work. These surveys can include questions about how AI tools are affecting productivity, innovation, and job satisfaction, offering insights into the strengths and weaknesses of the current AI strategy. Tracking **innovation metrics** is equally important. Leaders should monitor how many new AI initiatives are being proposed, how quickly AI projects are piloted and scaled, and the success rates of AI-driven experiments. These metrics offer a clear picture of how well the organization is leveraging AI to drive continuous improvement.

In addition to formal surveys and metrics, organizations should implement **continuous feedback mechanisms** that allow employees to provide input on AI tools, processes, and projects on a regular basis. These channels may include open forums, suggestion boxes, or dedicated AI feedback teams that report directly to leadership. Such mechanisms ensure that feedback is collected in real time and acted on promptly. This approach fosters a culture of open communication and empowers employees to actively contribute to the organization's AI journey.

Based on the insights gained from surveys, feedback, and performance metrics, leaders should be prepared to **adapt the AI strategy** as needed. If employees express resistance to certain AI tools, leadership should offer additional training or support to help them become more comfortable with the technology. If innovation metrics show a decline in AI-driven projects, leaders may need to introduce new incentives for creative exploration, such as hackathons or innovation challenges. Flexibility is key—adjusting the AI strategy in response to real-time feedback ensures that the organization remains agile and innovative in its AI endeavors.

Leadership's Role in Sustaining an AI-First Culture

Sustaining an AI-first culture requires strong leadership at all levels of the organization. Leaders must not only support AI initiatives but also take an active role in shaping the organization's AI vision, promoting continuous learning, and ensuring that AI practices align with the organization's values and goals. Leadership's direct involvement in AI efforts signals the importance of AI to the future of the business, inspiring employees to embrace the technology and fostering a culture that thrives on innovation and adaptability.

One of the most critical responsibilities for leadership is **leading by example**. When executives and managers use AI in their own decision-making processes, it demonstrates to employees that AI-driven insights can enhance business outcomes. This not only shows that AI is an integral part of the organization's future but also makes it easier for employees to follow suit. Leaders who embrace AI set the tone for how the rest of the organization will approach AI-driven projects, helping to remove fear or resistance toward adopting new technologies.

Equally important is **communicating AI's value**. Leadership must continuously explain how AI supports the organization's strategic goals and empowers employees to innovate and make more informed decisions. By articulating a clear vision of how AI will impact the business, leaders can align the entire workforce around common AI objectives. This helps employees understand that AI is not merely a technical tool but a driver of growth and competitive advantage.

In addition, leadership must ensure that the organization is equipped with the necessary resources by **investing in AI development**. This includes allocating funds for AI tools, infrastructure, and training programs that upskill employees in AI-related skills. Without these investments, AI initiatives are unlikely to achieve their full potential, and employees may struggle to keep up with the rapid pace of technological advancements.

Steps to sustain leadership in an AI-first culture include the following:

- **Champion AI initiatives:** Senior leaders should visibly and actively support AI projects. By being personally involved in key AI-driven initiatives, leaders can inspire employees to embrace AI adoption and participate in AI efforts.

- **Promote AI-driven decision-making:** Leaders must encourage the use of AI in decision-making across all levels of the organization. Providing managers with AI tools and ensuring that they integrate these tools into their workflows reinforces the importance of data-driven decision-making.

- **Foster a growth mindset:** Cultivate a culture of continuous learning within the leadership team itself. By staying informed about emerging AI technologies and encouraging experimentation, leaders set the example for adaptability and openness to new approaches.

For example, a multinational retail company's CEO became a vocal advocate for AI, discussing its potential to enhance customer experiences and streamline operations at internal meetings and public forums. The CEO led by example, using AI-powered dashboards to track key business metrics, which demonstrated the practical benefits of AI. This approach encouraged employees at all levels to adopt AI, resulting in a surge of AI-driven projects across the company that improved operational efficiency and customer satisfaction.

Key Takeaways for Leadership

To build and sustain an AI-first culture, leaders must take an active role in guiding their organizations through the process. Creating such a culture involves fostering innovation, collaboration, and continuous learning across all levels. Key recommendations for fostering an AI-first culture include the following:

- Articulate a clear AI vision that aligns with organizational goals and is consistently communicated to all employees.

- Encourage experimentation by creating safe spaces for innovation and rewarding employees who take AI-related risks and explore new ideas.

- Break down silos and promote collaboration across departments to ensure that AI projects benefit from diverse perspectives and expertise.

- Invest in continuous learning and upskilling to equip employees with the skills needed to succeed in an AI-driven environment.

- Promote ethical AI practices by prioritizing fairness, transparency, and data privacy.

- Lead by example by incorporating AI into decision-making processes and actively championing AI initiatives from the top down.

Driving Change and Accountability in AI Adoption

The successful implementation of AI within an organization is not just a technological endeavor—it requires visionary leadership that can drive change, foster alignment across departments, and ensure accountability at every stage of AI adoption. Whereas earlier chapters have addressed AI governance, culture, and workforce development, this chapter focuses on the leadership strategies necessary to move AI initiatives from vision to execution.

AI adoption often faces organizational resistance, misalignment between technical and business teams, and a lack of accountability for AI-driven outcomes. Leaders must bridge these gaps by not only championing AI initiatives but also creating the structures and incentives that ensure AI projects deliver measurable value. Without strong leadership, AI efforts can stall due to unclear ownership, ethical concerns, or failure to scale effectively.

This chapter explores the practical steps leaders must take to embed AI into daily business operations, drive organizational change, and maintain accountability for AI decisions. Unlike previous discussions on AI strategy and governance, this chapter provides a roadmap for leaders to actively guide their teams through AI adoption, ensuring that AI implementation is both effective and sustainable.

Translating AI Vision into Action: Leadership's Role in Driving Adoption

A compelling vision for AI is not just about articulating its potential—it is about embedding AI into the fabric of the organization in a way that drives real, measurable impact. Whereas previous discussions have addressed the strategic importance of AI, this section focuses on the active leadership role required to translate vision into execution, ensuring that AI adoption moves beyond abstract goals to become a sustained, organization-wide transformation.

To embed AI effectively, leaders must do more than just define AI's role in the company—they must actively demonstrate how AI integrates into daily decision-making, problem-solving, and innovation efforts. This means ensuring that AI is seen not as an isolated initiative but as a core part of how teams operate, collaborate, and create value. The key to success is aligning AI with business priorities in a way that is actionable, accessible, and adaptable.

Leadership also plays a crucial role in building confidence and trust in AI, especially in organizations where AI adoption may be met with skepticism. Employees must see AI as a tool that enhances their work rather than threatens their roles, and this requires ongoing, transparent communication from leadership. Leaders must address concerns head-on, providing clarity on AI's purpose, limitations, and benefits and ensuring that employees feel empowered rather than displaced by AI-driven changes.

Beyond setting the vision, leaders must foster an organizational mindset that embraces AI-driven innovation and learning. This requires not just communication but visible action: leaders should be directly involved in AI initiatives, sponsor AI-driven projects, and create spaces for experimentation, iteration, and cross-functional collaboration.

Momentum is built when AI delivers tangible wins that employees and stakeholders can see and feel. Leaders should showcase real-world examples of AI driving efficiency, improving customer experience, and enhancing operations. Sharing these success stories, whether through internal newsletters, company-wide meetings, or leadership briefings, helps transform AI from a conceptual initiative into a visible and trusted part of the company's evolution.

Ultimately, AI adoption is not a single decision—it is an ongoing, iterative process that requires constant reinforcement, adaptation, and engagement from leadership. By ensuring that AI is embedded into the organization's culture, workflows, and long-term strategy, leaders can create an environment where AI is not just accepted but actively embraced as a driver of growth, innovation, and competitive advantage.

Leading AI-Driven Change: Turning Resistance into Engagement

Successfully integrating AI into an organization requires more than technological upgrades—it demands a cultural transformation that reshapes workflows, decision-making, and employee mindsets. Although AI promises efficiency gains, improved decision-making, and new opportunities, it also introduces uncertainty, particularly among employees who may be concerned about job security or the disruption of familiar processes.

This section shifts the focus from why AI adoption is important to how leaders can navigate the resistance that naturally arises during AI transformation. Leaders must go beyond advocating for AI and actively manage the human side of the transition, ensuring that employees feel supported, informed, and engaged throughout the process.

One of the most significant leadership responsibilities in AI adoption is change management. AI is not just another technology implementation; it requires a fundamental shift in how work is performed and how decisions are made. Leaders must clearly define the role of AI, align AI initiatives with business goals, and proactively address employee concerns. If employees see AI as a mysterious, top-down initiative, resistance will grow. However, when AI is introduced transparently—with clear expectations, training, and open dialogue—employees are more likely to embrace its benefits.

Building trust is at the core of AI adoption. Employees are more receptive to AI when they feel involved in decision-making and understand how AI will affect their roles. Leaders must ensure that AI is positioned not as a replacement for human workers but as a tool to enhance their work. AI can automate repetitive tasks, allowing employees to focus on high-value activities such as strategic thinking, creativity, and problem-solving. Leadership must communicate what AI will change, what it will not change, and what opportunities it creates for employees to develop new skills and take on more meaningful work.

An effective way to reduce fear and skepticism is to directly involve employees in AI implementation. Instead of rolling out AI solutions as completed, top-down initiatives, organizations should engage employees in testing, feedback loops, and pilot programs. Employees who actively participate in AI projects are more likely to see AI's value firsthand, becoming advocates rather than skeptics.

For example, a manufacturing company implementing AI-driven predictive maintenance faced initial pushback from factory workers who feared that AI would replace their expertise. Instead of forcing AI adoption, leadership invited employees to collaborate with data scientists and engineers in training the AI system. Employees provided real-world insights into machine behavior, refining AI predictions and making them more effective. Over time, workers saw how AI enhanced their ability to prevent equipment failures, reduced unexpected downtime, and improved efficiency, leading to greater trust in AI and a willingness to integrate it into daily workflows.

To further ease the transition, leaders must implement structured AI training and skill development programs. Employees who feel unprepared for AI-driven changes may resist them, whereas those who receive adequate training gain confidence in using AI tools and applying AI-driven insights to their work. Leadership should provide hands-on training, mentorship programs, and continuous learning opportunities to ensure that employees remain engaged and capable in an AI-powered work environment.

Finally, leaders must communicate AI's impact consistently and transparently. Establishing regular updates, open forums, and Q&A sessions allows employees to voice concerns and gain clarity about AI's role in the organization. Rather than presenting AI as a finished product, leadership should encourage ongoing discussions about its implementation, evolution, and ethical considerations, reinforcing that AI adoption is a collaborative process rather than a corporate directive.

AI-driven transformation is not just about upgrading technology; it is about empowering people to thrive in an AI-augmented future. Leaders who manage change effectively, address concerns openly, and actively involve employees in AI initiatives can turn resistance into engagement, ensuring that AI is viewed as a valuable tool for innovation, productivity, and career growth rather than a threat.

By fostering trust, providing structured training, and embedding AI into workflows through employee participation, leaders can create a culture where AI is embraced as an opportunity rather than feared as a disruption. AI adoption is not just about automation—it's about collaboration, human–AI synergy, and ensuring that technology works for people, not against them.

Leadership's Role in Responsible AI Deployment

As organizations scale AI initiatives, ensuring that accountability becomes a critical priority. AI is no longer an isolated tool—it plays a central role in shaping business decisions, customer experiences, and ethical standards. Without proper oversight, AI systems can introduce unintended risks, such as biased decision-making, security vulnerabilities, and regulatory noncompliance. Leaders must establish governance structures that ensure AI is developed, deployed, and monitored in a responsible and ethical manner.

A key leadership responsibility in AI deployment is defining clear governance mechanisms that regulate AI use across the organization. This means creating policies that outline who is responsible for managing AI systems, how AI models should be monitored, and what safeguards must be in place to ensure compliance with ethical and legal standards. Governance structures serve as guardrails, ensuring that AI is used in alignment with business objectives while minimizing risks.

AI governance requires cross-functional collaboration. Leaders should establish an AI governance committee comprising IT, legal, data science, compliance, and business unit representatives. This committee provides a holistic oversight function, ensuring that AI-driven decisions remain fair, explainable, and aligned with company values. It also helps bridge gaps between technical teams and business leaders, ensuring that AI development and deployment stay connected to real-world impact.

AI's power comes with inherent risks. Poorly managed AI systems can introduce bias, violate data privacy regulations, or make decisions that lack transparency. Leaders must proactively manage these risks by implementing robust bias detection protocols, security measures, and legal compliance reviews before AI systems go live.

A strong risk management strategy includes

- **Bias mitigation:** AI models must be audited regularly to identify and eliminate unfair biases, ensuring that they do not disproportionately affect certain user groups.

- **Cybersecurity and data protection:** AI systems must be safeguarded against security breaches that could expose sensitive business or customer data.

- **Regulatory compliance:** AI use must align with global and industry-specific regulations, such as GDPR, CCPA, and industry ethics guidelines, to avoid legal repercussions and maintain public trust.

By embedding these safeguards into AI governance structures, leaders not only protect the organization from financial and reputational risks but also create AI systems that are trustworthy, explainable, and accountable.

Beyond governance and risk management, leaders must ensure that AI projects are delivering measurable business value. Accountability means tracking AI's performance not just in terms of technical accuracy but also in terms of its broader impact on the organization.

Key areas to assess include

- **Technical performance:** Is the AI model accurate, reliable, and functioning as expected?

- **Business impact:** Is AI improving efficiency, reducing costs, or driving revenue growth?

- **Ethical integrity:** Are AI decisions explainable and aligned with the company's values?

Regular AI auditing and monitoring allow leadership to detect performance issues early and make necessary adjustments before problems escalate. AI audits should be an ongoing process, not a one-time evaluation. By continuously tracking AI outcomes, organizations can ensure that their AI systems remain effective, unbiased, and aligned with business goals.

Building a Culture of AI Responsibility

Accountability in AI is not just about policies—it's about creating a culture where AI responsibility is ingrained at every level of the organization. Leaders must ensure that all employees, from executives to frontline staff, understand how AI impacts their work and how they can contribute to ethical AI use.

To foster this culture, leaders should

- Communicate AI governance policies clearly so that employees understand expectations around AI use.

- Encourage transparency by making AI-driven decisions explainable to both internal teams and external stakeholders.

- Empower teams with AI literacy to ensure that employees across departments understand AI's role in business operations and can use AI responsibly.

When AI accountability becomes an integral part of the organization's culture, businesses can scale AI responsibly while maintaining ethical integrity, operational excellence, and regulatory compliance.

Leadership's Role in Ensuring AI Accountability

AI is a transformative force, but without strong leadership, it can introduce risks rather than opportunities. By implementing governance structures, proactively managing AI risks, and defining accountability metrics, leaders can ensure that AI delivers real business value while operating within ethical and regulatory boundaries.

When leaders champion responsible AI adoption, they build trust among employees, customers, and stakeholders, ensuring that AI serves as a tool for progress rather than a source of uncertainty. AI accountability is not just about oversight; it's about embedding responsibility into every stage of AI's development and deployment so that organizations can confidently embrace AI's full potential.

Advancing Ethical AI and Sustainable Innovation

As AI becomes deeply embedded in business operations, leaders must ensure that it is deployed responsibly, ethically, and in alignment with long-term societal and business values. Unlike traditional governance concerns, AI introduces challenges related to autonomy, data reliance, and potential unintended consequences. This section expands on earlier discussions by exploring how organizations can embed ethical AI into their innovation strategies, ensuring that AI remains a force for positive transformation rather than a source of risk or distrust.

Ethical AI governance must go beyond regulatory compliance; it requires fostering a culture of responsibility, transparency, and trust across all AI initiatives. Leaders play a pivotal role in ensuring that AI decision-making aligns with fairness principles, stakeholder expectations, and long-term sustainability goals. Ethical AI can serve as a catalyst for competitive advantage, enabling organizations to innovate while mitigating bias, security risks, and unforeseen consequences.

AI's success depends on how well it is integrated into an organization's strategic vision. Ethical AI should be seen not merely as a risk-management function but as an opportunity to build trust, improve operational efficiency, and enhance decision-making. Companies that prioritize ethical AI can strengthen customer relationships, reduce regulatory risks, and create a workplace culture that supports responsible AI adoption.

To achieve this, leaders must establish ethical principles that guide AI projects from inception to deployment. These principles should not only reflect industry best practices but also align with the organization's mission and values. By embedding ethics into AI-driven strategies, companies can ensure that AI fosters inclusivity, enhances productivity, and contributes to long-term sustainability rather than becoming a liability.

Bias in AI systems remains one of the most significant ethical challenges, particularly because AI models learn from historical data, which may reflect existing inequalities or systemic biases. If left unchecked, biased AI can lead to discriminatory hiring practices, unfair lending decisions, or flawed medical diagnoses. Leaders must take a proactive stance in ensuring that AI models undergo rigorous fairness testing, bias detection, and continuous auditing before deployment.

Diverse and representative datasets must be used to prevent systemic bias from influencing AI decision-making. AI fairness assessments should be conducted regularly to identify disparities in outcomes. Real-time bias detection tools can be integrated into AI systems to flag and address potential discriminatory patterns before they affect business decisions.

Ensuring fairness in AI is particularly critical in industries such as manufacturing, finance, and healthcare, where AI influences hiring, workforce management, and quality control. Leadership must remain vigilant in overseeing AI's impact on different demographic groups, ensuring that it does not create unintended advantages or disadvantages.

For AI to gain widespread acceptance, organizations must ensure that AI-driven decisions are explainable. Many AI models, particularly deep learning and neural networks, operate as black boxes, making it difficult to interpret their decision-making processes. Lack of transparency creates distrust among employees, customers, and regulators, potentially limiting AI adoption.

Explainable AI (XAI) solutions help bridge this gap by making AI systems more interpretable. AI models should be designed to provide clear reasoning for their decisions, allowing end users to understand the logic behind recommendations. Employees should have the ability to verify AI-generated insights, ensuring that human oversight remains an integral part of the decision-making process.

Industries that rely on AI for critical functions—such as supply-chain management, predictive maintenance, and risk assessment—must prioritize AI transparency. Manufacturing companies using AI for quality control, for example, should implement systems that offer clear justifications for why a product is flagged as defective, enabling human workers to cross-check and validate AI recommendations.

AI ethics must be a core element of corporate governance rather than an isolated compliance function. Organizations should create AI ethics committees or governance boards to oversee AI implementation, ensuring that ethical considerations are embedded at every stage of AI development.

Leaders should focus on monitoring AI decision-making across business units, making sure AI remains aligned with strategic objectives. Ethical risk assessments should be conducted before deploying AI models to prevent unintended negative consequences. Escalation pathways must be established to address ethical concerns, allowing employees and external stakeholders to raise questions about AI applications.

By embedding AI governance into the company's broader corporate structure, organizations can ensure long-term accountability while fostering AI's positive impact on business operations.

Although AI is often viewed as a technology-driven tool, its real value lies in its ability to build trust—both within organizations and in external relationships. Ethical AI practices strengthen customer confidence, improve employee engagement, and ensure compliance with evolving regulations. Leaders must recognize that AI's success depends on how well it aligns with human values and business integrity.

By integrating ethical AI principles into business strategy, organizations can establish themselves as industry leaders, demonstrating that innovation and responsibility are not mutually exclusive. Ethical AI is not a constraint on progress; rather, it serves as the foundation for sustainable, transparent, and trustworthy AI adoption.

Instead of seeing ethics as a burden, businesses should embrace it as a means of unlocking AI's full potential. Responsible AI fosters innovation, enhances decision-making, and ensures that AI-driven transformation benefits both businesses and the communities they serve.

Leading by Example and Cultivating an AI-First Mindset

Adopting an AI-first mindset requires more than just strategic advocacy from leadership: it demands active engagement, structured implementation, and a commitment to embedding AI into daily business operations. Leaders must move beyond conceptual support and actively integrate AI into their workflows, ensuring that its benefits are realized across every level of the organization.

One of the most effective ways to drive AI adoption is for leaders to use AI tools in their own decision-making and workflow optimization. When leaders leverage AI for insights, automation, and strategic planning, they demonstrate its real-world benefits, encouraging employees to follow suit. This hands-on approach removes skepticism by showcasing AI as a practical enabler of efficiency rather than a theoretical concept.

Leaders should incorporate AI-driven analytics into performance monitoring, forecasting, and operational decision-making. Instead of relying solely on traditional reports, they should use AI-powered platforms to track business performance in real time, optimize resource allocation, and anticipate industry trends. AI-generated insights provide greater precision, allowing organizations to make proactive, data-driven decisions rather than reacting to challenges as they arise.

Beyond decision-making, workflow automation is another critical area of AI integration. Leaders should embrace AI-driven tools for meeting scheduling, document summarization, and knowledge management, illustrating how AI can reduce repetitive workloads and allow professionals to focus on higher-value initiatives. When leadership actively demonstrates AI's ability to enhance efficiency, employees are more likely to see AI as a tool that improves—not replaces—their roles.

To successfully integrate AI across an organization, leaders must foster an environment where employees feel empowered to experiment with AI solutions. This means creating structured opportunities for teams to test and refine AI applications in real-world scenarios. AI pilot programs allow departments to explore AI's capabilities on a small scale before committing to full implementation, minimizing risk while maximizing learning.

Cross-functional collaboration further strengthens AI innovation. AI adoption should not be limited to technical teams but should involve various business functions, ensuring that AI solutions align with actual operational needs. When AI engineers work alongside finance, marketing, and supply-chain teams, they gain a broader understanding of business priorities, leading to AI applications that are more practical and impactful.

Leaders should also allocate resources for AI experimentation. Providing employees with access to AI tools, training, and data enables them to explore new solutions and improve existing processes. AI innovation flourishes in organizations that remove bureaucratic barriers and actively support teams in testing AI-driven ideas.

For AI adoption to be sustainable, organizations must build AI literacy across all levels of the workforce. Employees who understand AI's capabilities and limitations are better equipped to integrate AI into their roles effectively. Leadership must take an active role in upskilling teams, ensuring that AI is not confined to specialists but becomes a fundamental competency across the organization.

AI training should be tailored to different roles, ensuring that employees gain the specific knowledge they need to work with AI tools. Introductory AI education should focus on the fundamentals: how AI systems work, where they add value, and how to interpret AI-generated insights. More advanced training should be available for those working closely with AI models, enabling them to refine AI-driven processes and collaborate effectively with data scientists.

Mentorship and peer learning further accelerate AI knowledge transfer. Pairing experienced AI users with employees less familiar with AI allows for hands-on guidance and real-world application. Organizations that embed AI education into professional development programs ensure that AI competency grows organically rather than being treated as an isolated initiative.

AI should not be an afterthought in decision-making—it must be integrated into business strategies from the outset. Leaders should incorporate AI-driven forecasting models into strategic planning, ensuring that decisions are based on real-time insights rather than outdated reports. AI can enhance operational reviews, helping organizations track performance, identify inefficiencies, and optimize processes with precision.

To achieve this, AI-generated insights should be part of executive briefings, boardroom discussions, and operational planning meetings. Leadership teams should require AI-backed recommendations in major business decisions, demonstrating that AI is a core intelligence asset rather than an optional tool. AI should be used to challenge assumptions, refine strategies, and provide alternative solutions that may not have been considered through traditional analysis.

Many organizations still treat AI as a secondary function, with AI initiatives isolated within innovation teams or research labs. Leaders must eliminate this fragmentation by ensuring that AI is embedded into the company's core operations. AI should not be viewed as an optional experiment—it should be a requirement for optimizing business processes and enhancing competitive advantage.

Aligning AI initiatives with business-wide objectives ensures that AI directly contributes to key performance indicators. Leaders must move beyond isolated AI projects and integrate AI into fundamental processes such as customer engagement, supply-chain management, and financial forecasting. When AI is embedded in core operations, it shifts from being a technology experiment to an essential driver of business success.

For AI adoption to succeed, leaders must also create accountability mechanisms. Assigning clear responsibilities for AI oversight ensures that teams are aligned with governance standards and business goals. Regular reviews of AI's performance, impact, and ethical implications help maintain responsible AI use while ensuring that it continues to deliver measurable value.

By focusing on hands-on AI adoption, structured implementation, and workflow transformation, this chapter ensures that AI is not just a concept supported by leadership but a tool that actively enhances efficiency, decision-making, and innovation. Leaders must not just endorse AI adoption; they must lead by example, using AI to optimize their own effectiveness and proving its value across the organization.

An AI-first culture does not emerge through passive support—it requires leaders to take an active role in using, promoting, and integrating AI into business processes. AI should not be seen as a tool for technical teams alone but as a foundational capability for all decision-makers.

When leadership embeds AI into their workflows, encourages experimentation, and invests in AI literacy, they create an environment where AI is not just an initiative but a core driver of business success. Leaders must transform AI from a concept into an operational reality, ensuring that their organizations are not just AI-ready but AI-embedded.

Steps to Lead by Example

Leaders can set the standard for AI adoption by integrating AI into their leadership decisions. Using AI tools to analyze data, generate insights, and make informed decisions about strategy and operations demonstrates the practical value of AI. This not only improves decision-making accuracy but also shows employees how AI can be applied to real-world business challenges.

Additionally, leaders should promote a culture of experimentation by encouraging teams to test and pilot AI applications across the organization. Providing opportunities to explore new AI solutions and supporting innovative projects helps embed AI into the company's DNA. Whether through pilot programs or smaller-scale AI initiatives, experimentation can lead to breakthroughs in efficiency and creativity, inspiring more widespread AI adoption.

Finally, leaders need to support AI skills development by investing in continuous learning opportunities for employees. This can include providing access to AI tools, workshops, and online training programs that empower employees to become more comfortable and proficient with AI technologies. By building a workforce that is AI-capable, leaders not only ensure the success of current AI initiatives but also prepare the organization for future advancements in AI.

By leading through example, promoting innovation, and investing in AI skills development, leaders can effectively foster an AI-first culture that drives the organization toward innovation, efficiency, and long-term success.

An example of leadership failing in AI adoption can be seen in a global manufacturing company that implemented AI-driven systems to optimize production processes and enhance supply-chain efficiency. The company adopted AI tools for predictive maintenance, inventory management, and real-time production tracking, expecting to reduce downtime and increase overall efficiency. However, leadership did not establish a robust governance framework to oversee how AI was integrated into existing operations or ensure accountability for AI-driven outcomes.

One of the most critical failures was the lack of oversight in the predictive maintenance system, which relied on AI models to predict when machinery would fail. Initially, the AI system seemed to improve maintenance schedules, but over time, discrepancies in data input and a lack of retraining the AI model led to inaccurate predictions. Leadership did not prioritize the ongoing auditing of AI performance, nor did they invest in training employees to identify and correct these issues. As a result, the AI began to overlook critical maintenance needs, leading to unexpected machine breakdowns and production delays.

In addition, the AI system for inventory management failed to account for sudden supply-chain disruptions and demand fluctuations. Because leadership did not involve key stakeholders in AI implementation—such as supply-chain managers and operations teams—the AI model was not fully aligned with the company's business realities. The result was poor inventory forecasting that led to stockouts in some regions and overstocking in others, increasing costs and customer dissatisfaction.

These failures led to significant operational disruptions, increased downtime, and financial losses. The company also faced damage to its reputation with clients due to delays in product delivery, which could have been mitigated with better leadership accountability. By not establishing proper governance structures, auditing mechanisms, or employee training, the leadership's lack of involvement caused the AI systems to underperform and create more problems than they solved.

This example underscores the importance of leadership in driving responsible AI adoption. Without clear oversight, cross-departmental collaboration, and accountability measures in place, AI implementations can lead to inefficiencies, operational risks, and negative impacts on the business that could have been prevented with more proactive leadership.

Key Takeaways for Leadership

Effective leadership is essential for the successful adoption of AI across an organization. Leaders must inspire a shared vision for AI, guide the organization through the necessary changes, ensure accountability, and promote ethical AI practices. By leading by example, fostering innovation, and maintaining transparency, leaders can drive AI adoption in ways that create long-term value for the organization.

Key recommendations for leadership include the following:

- **Inspire a clear AI vision** that aligns with the organization's strategic objectives and drives innovation across departments.

- **Champion organizational change** by addressing employee concerns and building trust through transparency and involvement in AI initiatives.

- **Establish governance frameworks** that ensure accountability for AI outcomes and maintain compliance with ethical and regulatory standards.

- **Promote ethical AI practices** by developing guidelines that address fairness, transparency, and bias mitigation.

- **Lead by example** by incorporating AI into leadership decision-making processes and encouraging a culture of experimentation and innovation.

Growing and Retaining AI Talent

As AI adoption accelerates across industries, organizations must prioritize the development and retention of AI professionals, including data scientists, machine learning engineers, and AI strategists. Attracting top talent is only the beginning; the real challenge lies in fostering an environment where AI professionals can grow, innovate, and remain committed to long-term organizational success. Leadership plays a crucial role in creating a workplace that not only values technical expertise but also supports continuous learning, career development, and meaningful contributions.

In this chapter, we'll explore how an effective AI talent strategy requires more than competitive salaries. AI professionals seek environments where they can work on cutting-edge projects, collaborate across disciplines, and contribute to initiatives that drive tangible business outcomes. Organizations that fail to provide these opportunities risk losing their most talented individuals to competitors that offer a more stimulating and rewarding work experience.

Leadership's Role in Nurturing AI Talent

AI professionals thrive in environments where they can experiment with new technologies, test emerging AI models, and push the boundaries of what AI can achieve. When organizations embrace a culture of innovation, employees feel empowered to take calculated risks and learn from both successes and failures. Leadership plays a key role in fostering this mindset, ensuring that AI teams are encouraged to explore emerging AI techniques without fear of failure. An organization that supports experimentation and continuous improvement will find it easier to retain AI talent.

Providing meaningful and impactful work is another crucial aspect of AI talent retention. AI professionals are drawn to roles that challenge their skills and allow them to solve complex problems. Leaders must ensure that AI teams contribute to high-impact initiatives that align with the company's long-term objectives. When AI professionals see their work driving real business value—whether through predictive maintenance, process automation, or customer experience enhancements—they develop a stronger connection to the organization. For example, a manufacturing company successfully retained its AI talent by integrating them into sustainability projects aimed at optimizing production efficiency and reducing energy consumption. Seeing their work contribute to corporate social responsibility initiatives gave AI professionals a greater sense of purpose, reinforcing their commitment to the organization.

Beyond meaningful work, mentorship and career development play a critical role in talent retention. AI professionals want to see a clear trajectory for career progression within their organization. Structured mentorship programs, executive coaching, and leadership training can provide AI employees with opportunities to refine their skills and transition into more strategic roles. Organizations that invest in mentorship ensure that AI talent remains engaged and motivated, as employees feel that they are continuously growing rather than stagnating in their roles.

Strategies for Growing AI Talent

Ensuring that AI professionals continue to evolve and expand their expertise is essential for sustaining innovation. Without growth opportunities, even the most talented individuals can become disengaged. Organizations must actively foster a culture of continuous learning, interdisciplinary collaboration, and career progression.

AI technology is evolving rapidly, and professionals in the field must stay ahead of the latest advancements. Leadership should prioritize ongoing education by funding AI certifications, sponsoring attendance at industry conferences, and forming partnerships with academic institutions to provide employees with access to cutting-edge research. Internal training programs can further support skill development, ensuring that employees remain knowledgeable about the latest AI techniques and best practices. By actively supporting lifelong learning, organizations demonstrate their commitment to employee growth while strengthening their internal AI capabilities. Figure 16-1 displays the end-to-end AI talent lifecycle, illustrating the key stages of managing AI professionals.

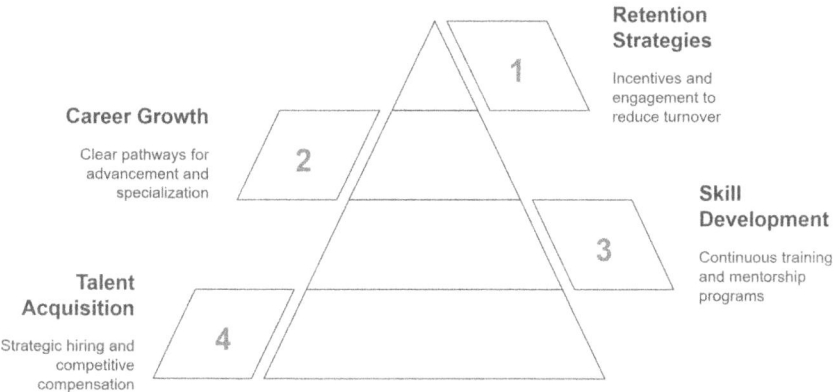

Figure 16-1: The AI talent lifecycle

Encouraging cross-functional collaboration also plays a critical role in growing AI talent. AI professionals often specialize in specific technical domains, but exposure to different business functions allows them to develop a broader understanding of how AI impacts various aspects of the organization. Leaders should create opportunities for AI teams to work alongside departments such as operations, supply chain management, finance, and product development. When AI professionals apply their expertise in different contexts, they not only sharpen their technical skills but also gain a deeper appreciation for business challenges, leading to more innovative solutions.

Career progression programs also ensure that AI professionals remain engaged and committed to the organization. Figure 16-2 highlights these key drivers behind effective AI talent retention. Organizations that provide structured development paths with increasing levels of responsibility and complexity create an environment where employees feel that their contributions are recognized and valued. A leading automotive company successfully implemented an internal AI career development program, allowing junior AI professionals to take on progressively challenging projects and offering senior AI experts leadership training. This structured approach helped create a strong pipeline of future AI leaders, reducing turnover and strengthening the company's AI capabilities.

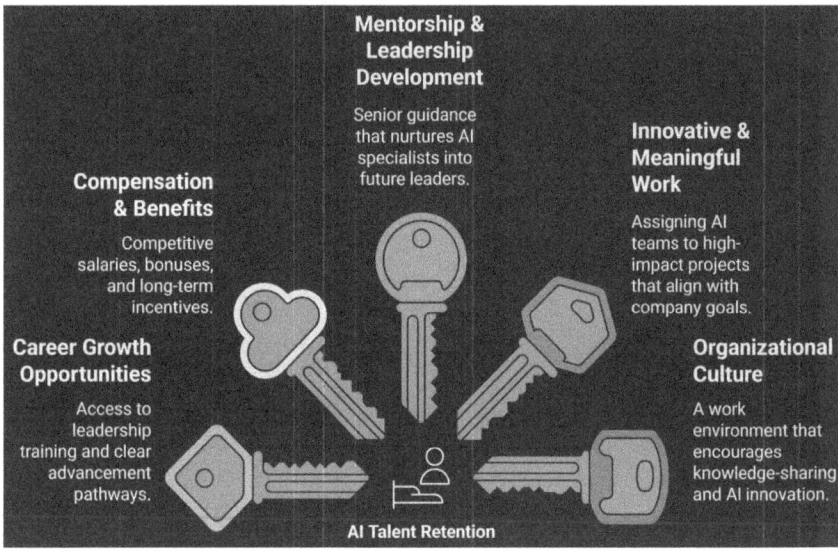

Figure 16-2: Factors that unlock AI talent retention

Retention Strategies: Keeping AI Talent Engaged and Motivated

Attracting top AI talent is only part of the challenge—retaining skilled professionals requires a long-term commitment from leadership to create an environment where AI employees feel valued, engaged, and motivated. Retention strategies must go beyond competitive salaries, focusing on career growth, work flexibility, and meaningful recognition. AI professionals thrive in environments that offer clear career progression, work–life balance, and opportunities for innovation. To sustain AI talent, organizations must ensure that employees see a future for themselves within the company and feel appreciated for their contributions. Table 16-1 offers a few solutions to some common retention struggles.

Table 16-1: Common AI Talent Retention Challenges and Solutions

CHALLENGE	SOLUTION
High demand and job-hopping	Provide long-term growth plans and leadership development
Burnout from project overload	Ensure manageable workloads and time for innovation
Lack of career visibility	Offer structured career roadmaps
Competitive job market	Strengthen organizational culture and AI engagement

A key component of retention is offering clear career progression paths. AI professionals are highly motivated by opportunities for advancement, and a lack of clear growth trajectories can lead them to seek opportunities elsewhere. Leadership must develop structured career pathways that allow AI employees to move into higher-level roles, whether that means becoming a lead data scientist, heading an AI strategy team, or transitioning into executive leadership. Providing a roadmap for career progression gives AI professionals a sense of direction and purpose, reinforcing their long-term commitment to the organization.

For example, a multinational electronics company implemented a tiered career progression system specifically designed for its AI professionals. Employees were given a clear set of milestones for advancement based on performance, project complexity, leadership capabilities, and contributions to AI innovation. This structured approach gave AI employees a transparent view of their potential career growth within the company, reducing uncertainty about their future and significantly improving retention rates.

Another critical factor in AI talent retention is offering flexible work arrangements. Given the nature of AI work—where problem-solving, research, and coding can often be done remotely—many AI professionals expect a degree of flexibility in how and where they work. Leadership should provide flexible work options, such as remote work capabilities, adjustable hours, and opportunities to work on passion projects within the company. By accommodating different work preferences, organizations create an environment where AI professionals feel trusted and empowered, leading to greater job satisfaction.

Beyond career growth and work flexibility, recognizing and rewarding AI innovation is essential for keeping AI professionals engaged. Many AI employees are driven by the impact of their work and want to feel that their contributions are making a difference. Leadership must implement recognition programs that highlight and reward AI-driven success, ensuring that employees feel valued for their efforts. Public recognition, financial incentives, and career advancement opportunities tied to innovation can all serve as powerful motivators.

For instance, a global tech company introduced an annual "AI Innovator Award" to recognize individuals and teams who contributed to innovative AI solutions that had a significant business impact. Employees who developed AI-driven efficiencies, advanced research, or launched new AI-powered products were celebrated in company-wide events and given financial bonuses. This not only boosted morale but also created healthy competition and motivation within the AI team, reinforcing the idea that AI contributions were both valued and rewarded.

Retaining AI talent requires leadership to be proactive in creating a fulfilling work environment that meets both professional and personal needs. By offering clear career progression, embracing work flexibility, and recognizing AI-driven innovation, organizations can build a culture where AI professionals remain engaged, motivated, and committed for the long term.

Leadership's Role in Building Loyalty and Trust

Building loyalty and trust between AI professionals and the organization is essential for long-term retention. AI specialists are highly sought-after, and if they feel undervalued or disconnected from the company's mission, they are more likely to seek opportunities elsewhere. Leadership plays a critical role in fostering an environment where AI talent feels appreciated, included in decision-making, and empowered to contribute to meaningful projects. Creating strong relationships between AI professionals and the organization requires open communication, a culture of collaboration, and leadership engagement in AI-driven initiatives.

Open communication and transparency are foundational to building trust. AI professionals want to understand how their work aligns with the company's strategic vision and how their contributions impact overall success. Leadership must ensure that AI teams are regularly updated on AI-related business goals, project outcomes, and long-term technological roadmaps. This can be achieved through internal newsletters, executive briefings, and company-wide forums where AI teams can share progress and insights. When leadership is transparent about the organization's AI objectives and challenges, AI professionals feel more connected to the company's mission and are more motivated to contribute to its success.

For example, a multinational corporation implementing AI-driven automation in its supply chain held quarterly town hall meetings where AI professionals could discuss their latest advancements, address roadblocks, and receive direct feedback from leadership. By integrating AI professionals into broader company discussions, the organization ensured that AI teams were not working in isolation but were actively engaged in the company's strategic direction. This openness reinforced trust and loyalty, making AI professionals feel like valued contributors rather than isolated technical specialists.

Beyond transparency, nurturing a collaborative leadership culture is key to fostering loyalty among AI professionals. AI talent thrives in environments where collaboration and innovation are championed at every level. Leaders must model this by being accessible and open to new ideas and actively supporting cross-functional teamwork. Creating platforms for AI professionals to engage directly with senior leadership fosters a sense of inclusion and reinforces the idea that AI innovation is a company-wide priority.

One effective approach is establishing leadership-led AI initiatives where AI professionals can work alongside executives and key decision-makers. For example, a global manufacturing firm created an AI Steering Committee led by senior executives; AI professionals were invited to present new ideas, propose AI-driven optimizations, and showcase ongoing AI projects. This initiative provided AI teams with direct access to leadership, allowing them to contribute to business strategy and innovation efforts. The result was an inclusive leadership culture where AI professionals felt valued, heard, and recognized for their expertise. This strengthened their connection to the company and increased their long-term commitment to its AI initiatives.

By fostering open communication, transparency, and collaborative leadership, organizations can build lasting trust and loyalty with their AI professionals. When AI teams see that leadership is invested in their success, values their contributions, and actively integrates AI into the organization's strategic vision, they are more likely to remain engaged and committed for the long term.

Case Study: Retaining AI Talent in a Manufacturing Company

A large manufacturing company had successfully implemented AI to optimize its production processes, but it faced challenges in retaining its top AI talent due to competition from tech companies offering more lucrative opportunities. To address this, leadership implemented several retention strategies focused on long-term career growth and job satisfaction.

The company introduced a comprehensive AI career development program that included opportunities for AI professionals to work on high-impact projects, access advanced training, and participate in leadership development workshops. Leadership also emphasized flexible work arrangements, allowing AI teams to work remotely and pursue passion projects aligned with the company's strategic goals.

Additionally, leadership launched an internal AI innovation recognition program, where top-performing AI professionals received bonuses and public acknowledgment for their contributions to business success. These initiatives fostered a sense of loyalty and commitment within the AI team, resulting in a 30% reduction in turnover rates and an increase in employee satisfaction scores.

Conclusion: Building and Retaining a Future-Ready AI Workforce

Hiring AI talent is just the beginning—long-term success depends on nurturing, engaging, and retaining these professionals. Leadership plays a critical role in fostering a work environment where AI teams feel valued, challenged, and aligned with organizational goals.

Continuous learning is essential. AI evolves rapidly, and leaders must invest in training, mentorship, and career development to keep talent engaged. AI professionals thrive in environments where they can innovate, collaborate, and contribute to impactful projects—organizations must provide meaningful work, clear career progression, and opportunities for experimentation.

Retention requires more than competitive salaries. Leaders should promote flexible work arrangements, recognition programs, and cross-department collaboration to keep AI professionals motivated. Open communication and transparency ensure that AI teams feel connected to the company's vision and understand their role in its success. Simply put, you can follow these key steps to support your AI workforce:

- **Invest in learning:** Provide ongoing AI training, mentorship, and career growth opportunities.

- **Foster innovation:** Encourage experimentation and support AI-driven projects.

- **Provide meaningful work:** Align AI talent with impactful, strategic initiatives.

- **Recognize and reward contributions:** Celebrate AI achievements through structured recognition programs.

- **Support flexibility:** Offer remote work and adaptable schedules to attract and retain top talent.

- **Encourage collaboration:** Break down silos, and integrate AI across business functions.

- **Ensure clear career paths:** Define structured progression opportunities for AI professionals.

- **Maintain transparency:** Keep AI teams informed and engaged in decision-making.

By prioritizing growth, engagement, and retention, leaders can build a strong, future-ready AI workforce that drives sustained innovation and success.

AI in Industry-Specific Applications: Customizing AI for Maximum Impact

AI's transformative potential is evident across all industries, but its true impact depends on how well it is customized to meet the specific challenges, regulations, and operational demands of each sector. Unlike generic AI applications, industry-specific AI solutions must address unique compliance requirements, market dynamics, and technological constraints.

In healthcare, AI must navigate stringent patient privacy laws while improving diagnostics and treatment planning. In finance, AI must balance real-time risk assessment with fraud detection while complying with financial regulations. In retail, AI-driven recommendation systems must align with shifting consumer behaviors while optimizing supply chain logistics. Each industry requires a distinct approach to AI implementation, ensuring that AI solutions not only drive efficiency but also align with sector-specific business models and compliance frameworks.

This chapter examines how AI is adapted and customized across industries, exploring the unique constraints, opportunities, and success stories of AI deployment in fields such as healthcare, finance, retail, manufacturing, and energy. By understanding how AI can be tailored to industry-specific needs, organizations can develop AI strategies that maximize efficiency, enhance decision-making, and drive long-term competitive advantages.

AI in Healthcare: Customizing AI to Meet Medical and Operational Demands

The healthcare industry presents unique opportunities for AI, but it also poses complex challenges that require industry-specific customization. Unlike AI applications in finance or retail, healthcare AI must comply with strict privacy regulations, integrate seamlessly with electronic health records (EHRs), and ensure explainability in medical decision-making. AI solutions in healthcare must be designed with a high degree of accuracy, reliability, and interpretability to support patient care without introducing risk.

One of the primary ways AI is customized for healthcare is in diagnostics and personalized medicine. AI-driven imaging tools, such as those analyzing X-rays or MRIs, must be trained on medically curated datasets and continuously validated by human radiologists to ensure accuracy. Unlike AI in other industries that can rely on probabilistic outputs, healthcare AI must provide high-confidence predictions that physicians can trust. Similarly, AI models for personalized medicine require integration with genetic databases, patient history, and real-world clinical data, allowing doctors to customize treatments for individual patients.

Beyond clinical applications, AI is being tailored to hospital operations and resource management. Unlike in other industries where AI can automate routine tasks without regulatory oversight, healthcare AI must balance automation with human intervention. Predictive models for hospital admissions, patient flow optimization, and surgical scheduling must be designed to work alongside medical staff, not replace decision-making. Furthermore, privacy constraints restrict how patient data can be used, requiring hospitals to implement federated learning models that allow AI to be trained on distributed datasets without compromising confidentiality.

To ensure compliance with healthcare regulations like HIPAA and the General Data Protection Regulation (GDPR), AI developers must build explainable models that provide auditable decision-making processes. Unlike AI in customer service or supply chain management, healthcare AI must be able to justify every decision, especially in high-stakes areas such as cancer detection or emergency medicine. Hospitals adopting AI must also invest in ongoing model validation, ensuring that AI systems do not degrade over time due to shifting patient demographics, emerging diseases, or changes in medical practice.

HEALTHCARE CASE STUDY

A major U.S. hospital system customized an AI-powered lung cancer detection tool to integrate with its existing radiology workflow. Although standard AI imaging tools were available, the hospital needed an FDA-approved, explainable AI model that could integrate directly into its PACS (Picture Archiving and Communication System) without disrupting radiologists' workflows. By training the AI on region-specific patient data and incorporating feedback loops with radiologists, the system reduced scan review times by 30% while ensuring that every AI-assisted diagnosis was auditable. The AI also reduced false negatives by 15%, leading to earlier cancer detection and improved patient outcomes.

AI in Finance: Enhancing Risk Management and Customer Experiences

The financial services industry has embraced AI as a powerful tool for automating complex processes, analyzing vast amounts of data, and improving decision-making. Financial institutions use AI to manage risks, detect fraud, and enhance customer engagement, but unlike other industries, financial AI must comply with strict regulatory frameworks while ensuring transparency and security. The ability to customize AI for specific financial applications is essential, as financial decision-making requires explainability, real-time processing, and heightened security measures.

AI has transformed fraud detection by allowing banks to analyze real-time transaction patterns and identify potential fraudulent activity. Traditional rule-based fraud detection methods often produced false positives, disrupting customer transactions. AI models, however, learn from transaction history, flagging only those that show significant anomalies. Financial institutions have integrated machine learning to detect fraud by recognizing subtle deviations in user behavior and flagging suspicious transactions before they are processed. AI's adaptability allows banks to refine fraud detection models continuously, responding to new tactics used by cybercriminals.

Risk management is another area where AI is driving significant transformation. Financial institutions rely on AI to assess and predict various risks, including credit risk, market volatility, and liquidity issues. Unlike traditional models that primarily assess risk based on financial statements and credit history, AI-driven risk management incorporates alternative data sources, including economic trends and customer behavior. AI models can simulate different economic conditions and predict potential downturns, allowing banks to adjust strategies accordingly. The ability to anticipate risks and respond proactively enhances financial stability and regulatory compliance.

AI is also improving personalized financial services. Chatbots and virtual financial advisors use AI to provide customers with tailored recommendations based on their financial habits. These AI-driven systems analyze transaction patterns to offer insights into budgeting, saving, and investing. Unlike generic financial advice, AI-driven solutions adapt to individual customer needs, ensuring a more relevant and engaging banking experience. This customization increases customer satisfaction while providing financial institutions with valuable insights into customer behavior.

Despite these advantages, financial AI faces significant challenges. Regulatory compliance remains a primary concern, as AI-powered decision-making must adhere to anti-money laundering (AML) regulations, know-your-customer (KYC) requirements, and data protection laws such as the GDPR. AI models must be designed to ensure transparency, allowing regulators and customers to understand how decisions are made.

Data security is another critical challenge. Financial data is highly sensitive, making it a prime target for cyber threats. AI-powered security solutions use behavioral analytics to detect potential breaches and unauthorized access, but financial institutions must also implement robust encryption, authentication, and access control measures. Unlike other industries where AI-driven data sharing is encouraged for collaboration, finance must prioritize data privacy and limit access to ensure compliance with strict security protocols.

FINANCIAL SERVICES CASE STUDY

A leading European bank successfully deployed an AI-driven fraud detection system to monitor transactions in real time. By leveraging machine learning, the system analyzed customer spending patterns and flagged suspicious activity before funds were transferred. The AI model adapted to new fraud schemes by continuously updating itself, reducing false positives and improving accuracy. As a result, fraudulent transactions decreased by

40% within the first year, saving the bank millions in potential losses. The AI system's transparency also ensured compliance with financial regulations, demonstrating the importance of explainable AI in finance.

AI's integration into finance is driving efficiency, security, and improved customer experiences. However, to fully realize AI's potential, financial institutions must customize AI solutions to address the industry's unique regulatory, security, and operational requirements. By leveraging AI for fraud detection, risk management, and personalized banking, financial organizations can enhance decision-making while maintaining compliance and protecting customer data. Customization ensures that AI solutions remain secure, ethical, and effective in navigating the complexities of the financial sector.

AI in Retail: Revolutionizing Customer Engagement and Supply Chain Management

AI is reshaping the retail industry by enabling businesses to better understand consumer behavior, personalize customer interactions, and optimize supply chain operations. Retailers that leverage AI can enhance customer experiences, improve demand forecasting, and streamline logistics. Unlike traditional retail strategies, AI-driven solutions analyze vast amounts of data in real time, allowing businesses to adapt quickly to changing consumer preferences and market conditions.

Retailers are using AI to revolutionize personalized marketing. AI-driven recommendation engines analyze customer browsing history, past purchases, and demographic data to offer tailored product suggestions. Unlike generic marketing strategies, AI ensures that each customer receives highly relevant recommendations, increasing conversion rates and improving customer satisfaction. Personalized marketing campaigns powered by AI also help retailers engage with consumers through dynamic content, such as AI-generated email promotions and targeted advertisements. The ability to customize recommendations in real time enhances customer loyalty, as shoppers receive a curated experience based on their preferences.

Demand forecasting has become significantly more accurate with AI. Traditional forecasting methods often struggle to adapt to sudden shifts in consumer demand, but AI models can analyze sales data, market trends, and external factors such as weather patterns and social media activity to predict demand fluctuations. By incorporating real-time data, AI enables retailers to optimize inventory levels, reducing instances of stockouts or overstock. This level of precision helps retailers manage supply chain costs while ensuring that popular products remain available to customers.

AI is also streamlining supply chain management by optimizing logistics and warehouse operations. Retailers face challenges in coordinating complex supply chains, particularly in e-commerce, where delivery speed and efficiency are crucial. AI-powered systems analyze logistics data to predict potential disruptions, such as transportation delays or supplier shortages. By using machine learning models, retailers can optimize delivery routes, manage warehouse inventory more efficiently, and automate order fulfillment. AI's predictive capabilities allow retailers to anticipate supply chain issues before they arise, reducing operational costs and improving order accuracy.

Despite these advantages, implementing AI in retail comes with challenges. Data privacy remains a significant concern, as retailers must ensure compliance with regulations such as the GDPR. AI-driven marketing relies heavily on consumer data, making it essential for retailers to maintain transparency about how customer information is collected and used. Unlike industries where data is used for purely operational purposes, retail AI must balance personalization with privacy to maintain consumer trust.

Another challenge is adapting AI to rapidly changing consumer preferences. Unlike static business models, retail trends evolve quickly, requiring AI systems to continuously update and refine their algorithms. AI models trained on historical data must be retrained frequently to reflect current shopping behaviors, ensuring that recommendations and demand forecasts remain relevant. Retailers that fail to update their AI systems risk making inaccurate predictions that could lead to lost sales or ineffective marketing campaigns.

RETAIL CASE STUDY

A global e-commerce company successfully deployed an AI-powered recommendation engine to enhance customer engagement. The AI system analyzed customer behavior in real time, predicting which products individuals were most likely to purchase based on their online activity. The deep learning algorithms adjusted recommendations dynamically, adapting to new browsing patterns and emerging consumer trends. As a result, the company experienced a 25% increase in average order value and higher customer satisfaction ratings. By personalizing the shopping experience, the retailer strengthened customer retention and improved sales performance.

AI is transforming retail by providing businesses with deeper insights into consumer behavior, optimizing inventory management, and enhancing logistics efficiency. To fully harness AI's potential, retailers must customize AI solutions to align with their business objectives, ensuring that AI-driven strategies enhance both customer experiences and operational effectiveness. By embracing AI for personalized marketing, demand forecasting, and supply chain optimization, retailers can stay competitive in an increasingly digital and customer-centric market.

AI in Energy: Optimizing Resource Utilization and Sustainability

The energy sector is undergoing a transformation as AI enables organizations to optimize resource usage, reduce operational costs, and enhance sustainability initiatives. With increasing global demand for energy and the shift toward renewable sources, AI is playing a crucial role in managing energy grids, predicting consumption patterns, and improving the efficiency of wind, solar, and other renewable energy sources. Unlike traditional energy management systems, AI can analyze vast amounts of real-time data, enabling predictive decision-making and automated optimization to enhance both efficiency and sustainability.

One of the most impactful applications of AI in the energy sector is energy grid management. Power grids are complex, with fluctuating demand levels that require constant adjustments to maintain stability. AI-powered systems analyze historical consumption data, real-time energy usage, and weather forecasts to predict demand surges and optimize energy distribution. By dynamically adjusting energy flow based on demand patterns, AI helps prevent blackouts, reduce energy waste, and improve grid reliability. This intelligent management of energy distribution allows utility providers to make data-driven decisions that improve overall efficiency.

AI is also playing a critical role in renewable energy optimization. Unlike fossil-fuel-based energy sources, wind and solar power depend on environmental factors that can be unpredictable. AI systems analyze real-time sensor data from wind turbines, solar panels, and energy storage units to predict and maximize energy output. For example, AI can assess wind speed, temperature, and sunlight exposure to adjust turbine blade angles or reposition solar panels for optimal performance. This level of automation increases the efficiency of renewable energy sources, ensuring that they contribute more effectively to the overall energy supply while reducing dependence on traditional power generation.

Beyond grid and renewable energy management, AI is enhancing energy consumption monitoring for both businesses and individual consumers. AI-powered systems track energy usage patterns, identify inefficiencies, and provide recommendations on how to reduce consumption. For businesses, AI can automate building energy management by adjusting heating, cooling, and lighting based on occupancy and usage trends. Consumers benefit from smart energy monitoring systems that provide real-time insights into household energy consumption, empowering them to make informed decisions about reducing their carbon footprint and energy costs.

Despite its advantages, implementing AI in the energy sector comes with challenges. One of the primary obstacles is managing the complexity of energy-related data. The energy industry generates massive amounts of data from sensors, smart meters, weather reports, and real-time consumption tracking. AI systems must be capable of processing and integrating this diverse data to generate accurate insights. Ensuring that AI models are trained on high-quality, representative datasets is essential to making reliable energy predictions and optimizations.

Another significant challenge is regulatory compliance. The energy sector is highly regulated, with strict requirements for reliability, security, and sustainability. AI-driven energy management systems must align with these regulatory standards while maintaining the integrity of energy supply. Utility companies and renewable energy providers must work closely with regulators to ensure that AI systems comply with energy policies, environmental standards, and cybersecurity regulations to prevent disruptions in service or legal challenges.

ENERGY SECTOR CASE STUDY

A European energy company successfully leveraged AI to enhance the performance of its wind energy farms. The company deployed an AI-powered system that analyzed real-time sensor data from wind turbines, including wind speed, temperature, and turbine performance metrics. The AI system continuously adjusted turbine settings based on weather conditions, ensuring that each turbine operated at peak efficiency. As a result, the company increased its overall energy output by 15% while reducing maintenance costs by 20%. By using AI-driven insights, the company was able to maximize renewable energy production and improve cost efficiency, demonstrating how AI can contribute to sustainable energy solutions.

AI is revolutionizing the energy industry by optimizing grid management, improving renewable energy performance, and enabling smarter energy consumption. However, for AI to be fully effective, energy providers must overcome data complexity challenges and ensure regulatory compliance. By integrating AI-driven solutions, energy companies can enhance efficiency, reduce environmental impact, and contribute to a more sustainable future. As the global energy landscape evolves, AI will continue to play a pivotal role in shaping a cleaner, more resilient energy system.

Key Takeaways for Leadership

AI's impact across industries is profound, but to realize its full potential, organizations must customize AI solutions to fit the unique needs and challenges of their sector. By tailoring AI technologies to industry-specific applications, companies can drive innovation, improve efficiency, and gain a competitive advantage.

Key recommendations for leadership include the following:

- **Identify industry-specific AI opportunities** that align with the organization's strategic goals, whether in healthcare, finance, retail, manufacturing, or energy.

- **Overcome challenges such as data privacy, regulatory compliance, and legacy system integration** by developing a clear AI strategy that addresses these issues.

- **Leverage AI to optimize operations, enhance customer experiences, and drive innovation** by focusing on applications that deliver the highest value for the industry.

- **Invest in upskilling the workforce** to ensure that employees are prepared to work alongside AI technologies and can contribute to AI-driven transformation.

Leading Organizational Change in the Age of AI

Artificial intelligence is not merely a tool—it's a transformational force that's redefining how manufacturers design, operate, and lead. For manufacturers to thrive in this new era, leadership must go beyond the technical implementation of AI and instead focus on creating a culture and structure that embraces AI as a core capability. Organizational change, when guided by human-centered leadership, becomes the key to integrating AI effectively and sustainably.

Rethinking Culture in an AI-Driven Environment

As AI becomes embedded in daily operations, it influences not only what work is done but also how people work together. Traditional top-down hierarchies often struggle to keep pace with the rapid, iterative cycles of AI development and deployment. In contrast, organizations that embrace an agile, collaborative, and learning-oriented culture are better positioned to succeed. Leaders must cultivate a culture that values data-informed decision-making, experimentation, and psychological safety.

In practice, this means creating an environment where employees are not just passive recipients of AI tools but active participants in

215

identifying where AI can improve processes. For instance, frontline workers and plant managers—who understand the nuances of daily operations—can offer invaluable insight into pain points and inefficiencies that AI might address. Inviting their input during pilot projects and AI testing can surface more practical use cases, drive stronger adoption, and increase trust.

Open innovation environments, such as internal AI labs or cross-functional innovation sprints, give employees the freedom to explore AI solutions and propose ideas without fear of failure. These open environments encourage experimentation and free innovative thinking. The ability to offer an environment like this is an important item that is overlooked by many companies. This cultural shift empowers teams to continuously improve while adapting to the evolving landscape of AI.

Building Trust and Addressing Resistance

Despite AI's promise, it often triggers resistance rooted in concerns about job loss, surveillance, or the opacity of algorithmic decision-making. To overcome this resistance, leaders must engage in transparent, ongoing communication. Explain why AI is being adopted, what changes it will bring, and how it will support—not replace—employees. Remember what was said about email and ATMs, but we still have the post office and bank tellers. Gen AI is no different: it is here to augment, not to replace.

For example, if an AI-based scheduling tool is implemented, leaders should highlight how it can reduce fatigue, improve fairness, and create better work–life balance—not just increase efficiency. Framing AI through the lens of human benefit makes adoption feel more personal and less threatening.

Reskilling is a vital part of building this trust. When employees see the company investing in their future—offering training in robotics, machine learning, or data interpretation—they're more likely to support transformation efforts. Upskilling programs should be tailored to the organization's AI roadmap, ensuring that the workforce evolves alongside the technology.

Identifying and empowering internal AI champions—early adopters who can model new behaviors and mentor peers—also accelerates cultural change. These individuals serve as bridges between technical teams and the broader workforce, translating AI's potential into relatable, job-specific outcomes.

Aligning Organizational Design with AI Strategy

Effectively scaling AI in manufacturing requires more than technology upgrades: it calls for a deliberate reshaping of organizational structures to enable cross-functional collaboration and system-wide alignment. Traditional departmental silos between IT, engineering, operations, and business strategy often hinder the integration of AI initiatives. For AI to move from isolated pilots to enterprise-wide impact, leaders must foster structures that enable teams to work together, share data, and solve problems holistically.

Manufacturing leaders are increasingly shifting toward integrated team models that break down barriers between departments. One approach involves embedding AI professionals—such as data scientists, machine learning engineers, and AI product managers—directly into existing operational or engineering teams. This model ensures that domain expertise and AI expertise work hand in hand from project inception to deployment. Alternatively, some manufacturers establish centralized AI centers of excellence or innovation hubs to serve as internal consultancies, guiding cross-departmental initiatives and knowledge sharing.

Regardless of structure, success hinges on strategic alignment. AI teams must be explicitly tasked with addressing business-critical goals such as increasing uptime, improving first-pass yield, and advancing sustainability metrics. When AI is linked directly to core performance indicators—not just treated as an experimental add-on—it earns organizational momentum. Employees are more likely to engage when they see AI solutions driving measurable improvements in their work environments, such as reducing maintenance emergencies or minimizing manual quality checks.

To reinforce this alignment, some organizations have introduced formal AI leadership roles—such as Chief AI Officers and AI Program Directors—who oversee AI strategy, coordinate initiatives across departments, and ensure that resources are allocated appropriately. These roles help avoid redundant efforts and fragmented projects while serving as champions of ethical and scalable AI practices. In addition, AI leadership can mediate between the technical and business sides of the organization, translating priorities in both directions to keep initiatives moving forward.

AI tools themselves can also support better organizational design. Advanced project management platforms now incorporate AI capabilities to analyze team workflows, identify bottlenecks, and offer data-driven recommendations to improve efficiency. For example, AI can assess how handoffs between departments slow product development cycles or how resource allocation impacts factory throughput. These insights help leaders redesign cross-functional collaboration models in ways that would be difficult to spot manually.

When forming cross-functional AI teams—whether for product design, supply chain optimization, or predictive maintenance—it's critically important to clearly define roles and responsibilities from the outset. This avoids confusion and ensures accountability across disciplines. AI practitioners need to understand how their work ties into production or engineering needs, just as operational staff must understand how AI insights will inform decision-making. Setting expectations early builds trust and reduces friction, especially when projects span departments with different priorities and vocabularies.

Equally important is establishing a neutral, shared space for collaboration. Using digital platforms like Microsoft Teams, Trello, Confluence, and Slack helps unify communication and document progress. These tools provide a common ground where all stakeholders—whether from IT, operations, quality, or leadership—can share updates, troubleshoot, and brainstorm solutions together. Having a digital "command center" also reduces the perception that any one group is dominating the conversation or "owning" the initiative, which can be especially valuable in organizations with strong departmental identities.

Although shared platforms promote cohesion, training and communication strategies must remain tailored. For example, training operators on how to interpret AI-driven maintenance alerts should differ from onboarding engineers into a computer vision project. Department-specific training respects context and keeps engagement high. It avoids the feeling that AI is being imposed from outside, which can breed resistance. Instead, employees are empowered to adopt AI tools that make sense for their day-to-day responsibilities.

Ultimately, successful alignment of AI strategy with organizational design relies on leadership's ability to orchestrate structure, roles, tools, and culture into a cohesive system. It requires not just vision but execution—putting in place the processes, support structures, and shared language that enable people and technology to work together at scale. As manufacturing enters the era of pervasive AI, it's this orchestration that will distinguish organizations that scale effectively from those that stall in perpetual pilot mode.

Empowering the Workforce for AI Adoption

Empowerment means more than training—it's about embedding AI literacy into the fabric of the organization. Leaders should ensure that employees at all levels understand what AI is, what it does, and how it supports their specific role. This can include foundational courses on AI concepts, hands-on workshops with AI tools, and role-based learning paths.

Executives must also lead by example. When senior leaders use AI dashboards for planning or share insights from AI analysis during decision-making, it signals that AI is central to business operations. Their engagement reinforces a culture where curiosity, collaboration, and learning are expected—and rewarded.

Crucially, workforce empowerment includes establishing feedback loops. AI systems can behave unpredictably or generate outcomes that don't align with on-the-ground realities. Think of this as a hallucination: the system does not have a direct answer, so it creates its own answer that may not be rooted in actual fact. Line workers are often the first to spot such issues. Building formal mechanisms for employees to report problems, suggest adjustments, or participate in system improvements makes AI more resilient and human-centered.

In one real-world example, a manufacturing plant implemented AI to predict equipment failures. Operators noticed that the AI was missing signals that they had learned to recognize intuitively. When leadership incorporated operator feedback into the model's retraining, both performance and trust improved significantly.

Addressing the Psychological Impact of AI Adoption

AI-driven organizational change doesn't just present logistical and technical challenges, it also introduces significant psychological factors that influence employee reactions. Many employees experience fear, uncertainty, or skepticism when faced with AI adoption, which can slow transformation efforts. Understanding and addressing these emotional responses is crucial to managing resistance effectively.

The **Kubler–Ross Change Curve**, originally developed to describe emotional responses to grief, is often applied to workplace change, including AI integration. Figure 18-1 outlines these five stages that employees may experience:

1. **Denial:** Employees resist AI, believing it will not impact their roles significantly.

2. **Anger:** Fear of job loss or disruption may lead to resentment toward AI initiatives.

3. **Bargaining:** Employees seek reassurances or compromises regarding AI's role in their work.

4. **Depression:** A sense of helplessness can set in if employees feel unprepared for AI-driven changes.

5. **Acceptance:** Employees begin to embrace AI and recognize its potential benefits.

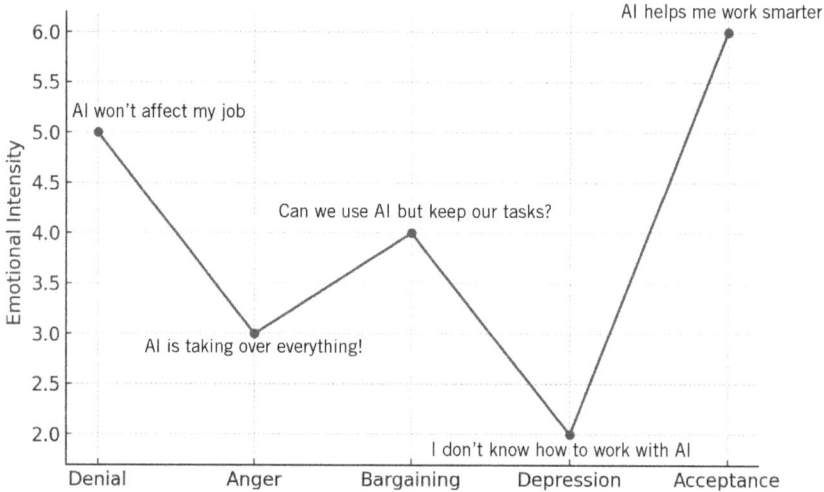

Figure 18-1: Kubler–Ross Change Curve for AI adoption

Organizations can mitigate negative reactions by supporting employees throughout this curve with transparent communication, leadership engagement, and reskilling programs. Providing employees with AI coaching sessions, peer support groups, and structured change management strategies helps ease the transition.

Another effective model is the ADKAR Change Model, which breaks down change adoption into five steps: awareness, desire, knowledge, ability, and reinforcement (Hiatt, 2006). Organizations implementing AI should ensure that employees

- Understand why AI is being introduced (**awareness**)
- Are motivated to engage with AI rather than resist it (**desire**)
- Have access to training programs and AI literacy initiatives (**knowledge**)
- Can apply new AI-related skills in their day-to-day roles (**ability**)
- Receive continuous support and reinforcement from leadership (**reinforcement**)

By recognizing the emotional and psychological barriers to AI adoption and proactively addressing them, organizations can create an environment where employees feel empowered rather than threatened by AI-driven change.

Leading with Purpose in the AI Era

Ultimately, AI transformation is not a technology problem: it is a leadership imperative. To succeed, manufacturing leaders must connect AI adoption to the organization's mission and values. AI should be framed not just as a cost-saving tool but also as a force for good: improving worker safety, reducing environmental impact, increasing agility, and unlocking innovation.

Purpose-driven leadership involves making ethical decisions about AI use, ensuring that systems are fair, transparent, and accountable. It means considering how AI affects communities, customers, and the broader ecosystem. It also requires empathy—understanding the emotional and psychological impacts of change, and supporting employees through uncertainty.

As manufacturing enters the era of intelligent operations, leaders must balance speed and responsibility, as well as ambition and empathy. Those who can unite vision with values will not only deliver smarter factories but also stronger, more resilient organizations. In the age of AI, the future of manufacturing leadership lies not in managing machines but in empowering people to thrive alongside them.

AI-Driven Innovation: Unlocking New Business Models and Opportunities

The integration of AI into business processes has moved beyond automation and efficiency improvements—AI is now a critical enabler of innovation, transforming industries and unlocking entirely new business models. Organizations that embrace AI not only optimize existing operations but also position themselves to create disruptive products, services, and strategies that were previously unimaginable. By leveraging AI, businesses can enter new markets, develop customer-centric solutions, and discover fresh opportunities for growth.

This chapter explores how AI is driving innovation, highlights real-world examples of AI-driven business models, and outlines strategies for fostering an innovation mindset within organizations. We will also look at how companies can identify and capitalize on AI-driven opportunities to gain a competitive edge.

The Role of AI in Driving Innovation

AI's capacity to analyze large datasets, uncover patterns, and predict outcomes makes it an essential tool for fostering innovation across industries. By streamlining the product development process, enabling personalized customer experiences, and driving the creation of new business models, AI helps organizations identify and seize previously unrecognized opportunities. This AI-powered innovation allows businesses to adapt quickly to shifting market conditions, cater to evolving customer needs, and stay ahead of their competition.

AI empowers organizations to develop entirely new products and services that leverage machine learning and automation. For instance, AI-driven platforms can offer personalized financial advice, predictive healthcare solutions, or intelligent supply chain management tools. Additionally, AI enhances personalization, allowing businesses to deliver highly targeted products, services, and marketing to individual customers. This level of hyper-personalization not only strengthens customer engagement and loyalty but also drives significant revenue growth.

AI also unlocks new business models by transforming how companies operate and generate revenue. Organizations can shift from selling traditional products to offering AI-powered services, adopt data-driven subscription models, or create AI-driven marketplaces that connect customers and suppliers in innovative ways. This shift opens new revenue streams and helps companies stay competitive in rapidly evolving markets.

To harness AI-driven innovation, organizations should first identify high-impact areas where AI can bring the most value, such as improving customer experiences, enhancing product development, or optimizing operations. Encouraging a culture of experimentation is also critical— teams should be given the freedom to explore AI's potential through pilot projects and innovation sprints that test new AI-powered solutions. Finally, leveraging data as a strategic asset is crucial for fueling AI-driven innovation. Companies must invest in robust data infrastructure to ensure data is collected, processed, and analyzed effectively, supporting AI's ability to drive new growth opportunities.

AI-Powered Business Models

AI is enabling businesses to rethink traditional models and develop innovative approaches that leverage AI's unique capabilities. From AI-as-a-Service (AIaaS) to intelligent marketplaces, these AI-driven models provide scalable, flexible, and personalized solutions that cater to evolving customer needs.

One prominent AI-powered business model is AIaaS, where companies offer AI capabilities as a service, allowing businesses to integrate machine learning, natural language processing, and data analytics into their operations without building AI infrastructure in-house. This model provides scalability and flexibility, making AI adoption more accessible and efficient.

AI-driven marketplaces are revolutionizing how buyers and sellers connect in the manufacturing ecosystem. These platforms use machine learning to optimize pricing, forecast demand, and recommend products and suppliers based on historical patterns. For manufacturers, these marketplaces can streamline procurement, reduce costs, and unlock new sales channels. For example, AI-powered B2B platforms can automatically match component buyers with qualified suppliers based on lead time, quality scores, and regional availability.

To take advantage of these platforms, manufacturers should follow a few high-level steps:

1. **Identify use cases:** Start by mapping business processes where marketplace dynamics play a role, such as sourcing raw materials, selling surplus inventory, or distributing finished goods. Determine which functions could benefit from AI-enhanced discovery, pricing, or matching.

2. **Explore available platforms:** Research industry-specific AI-powered marketplaces. Examples include Thomasnet and Xometry (for custom parts and supplier discovery), Alibaba and Amazon Business (for general industrial products), and even vertical AI marketplaces like Graphite and Turing (for digital services). Evaluate each platform's capabilities: does it offer AI-based recommendations, predictive pricing, and intelligent negotiation features?

3. **Assess integration potential:** Consider how a marketplace can integrate into your existing systems, such as enterprise resource planning (ERP), manufacturing execution systems (MES), and procurement tools. Many platforms offer APIs or prebuilt connectors to enable smoother adoption. Start small by piloting with a single product category or supplier relationship.

4. **Define a data strategy:** AI marketplaces rely on data—your pricing history, demand forecasts, supply chain performance—to deliver insights. Ensure that your internal data is clean, accessible, and ready to feed into these platforms. Some providers offer onboarding support and data mapping services to ease this process.

5. **Monitor and iterate:** Once integrated, track key metrics such as procurement efficiency, price competitiveness, and lead-time improvements. Use this feedback loop to refine strategies—like adjusting bidding thresholds and targeting new regions.

In parallel, organizations should evaluate their broader business model to explore where AI adds value. This could include offering products "as a service," using AI for dynamic pricing models, or providing AI-enhanced customer portals for configuration and ordering. By embedding AI into these touchpoints, companies can not only improve internal operations but also open new revenue streams and customer experiences.

Ultimately, AI-powered marketplaces are not just tools—they are ecosystems. Manufacturing leaders who adopt them early and strategically will gain speed, visibility, and competitive edge in a fast-evolving landscape.

Identifying AI-Driven Opportunities for Growth

AI holds enormous potential to fuel growth by revealing hidden patterns, uncovering customer needs, and spotting emerging trends across vast datasets. For manufacturers, this means the ability to launch new products faster, expand into untapped markets, optimize product–market fit, and better serve customers. But to realize these gains, manufacturers must move beyond surface-level insights and adopt a structured approach to leveraging AI in market research, customer behavior analysis, and strategic forecasting.

At its core, AI thrives on identifying relationships in complex, multivariable datasets—something humans cannot easily do at scale. For instance, natural language processing (NLP) can process thousands of online reviews, support tickets, and social media comments to detect recurring pain points and preferences in product features. Clustering algorithms can segment customer behavior into distinct groups based on purchasing patterns, usage data, and channel interactions, revealing which customer types are underserved.

Let's say a midsized manufacturer uses AI to analyze support call logs and after-sales service data. NLP tools surface that customers in colder regions are disproportionately requesting service for a specific product line in winter months. This insight leads the company to develop a weather-resistant upgrade and reposition the product in cold-climate markets—resulting in a 15% boost in seasonal sales.

Similarly, machine learning models trained on sales data, competitive pricing, and macroeconomic indicators can generate demand forecasts and simulate market expansion scenarios. For example, an AI platform may recommend launching a lower-cost variant of a product in Southeast Asia after detecting high price sensitivity and favorable import conditions. AI doesn't just highlight *what* is happening; it helps predict *what's next*.

To start using AI for market research, manufacturers can take the following high-level steps:

1. **Aggregate relevant data:** Pull internal and external data sources into one place—this may include sales reports, CRM records, online reviews, market reports, and competitor websites. Tools like Crayon and AlphaSense can help with competitive monitoring, and platforms like Talkwalker and Brandwatch are used for AI-powered sentiment analysis and social listening.

2. **Use prebuilt AI tools:** You don't need a large data science team to get started. Platforms like MonkeyLearn, Tableau with Einstein AI, and Google Cloud's AutoML offer plug-and-play AI capabilities for text classification, trend detection, and forecasting.

3. **Spot market gaps:** Feed customer complaints, feature requests, and sales anomalies into clustering algorithms and topic modeling engines. These help identify product deficiencies and underserved customer segments that could become growth opportunities.

4. **Test and iterate:** Run small pilot programs to validate AI insights. If a model suggests high demand for a feature in a new region, test a regional campaign or offer a localized variant. Measure conversion rates, feedback, and retention before scaling further.

AI IN PRODUCT INNOVATION

Consider a global tools manufacturer that wanted to expand into emerging markets. Using AI to analyze customer service transcripts, the company found a recurring complaint: power drills frequently failed in regions with unstable voltage. This insight—missed by traditional reporting—led to the development of a voltage-regulating variant. Launched through an AI-informed pricing model tailored to local income levels, the product captured significant market share in its first year.

Another company used generative AI to simulate customer personas based on historical sales and service data. These synthetic personas were then used to test different product concepts and marketing messages through an AI-powered A/B testing platform. The resulting insights increased lead conversion by 22% across B2B buyers.

Scaling AI-Driven Innovation Across the Organization

Once an organization begins unlocking AI-driven innovation, the next crucial step is scaling these innovations across the business. Scaling AI involves moving beyond pilot projects and small-scale initiatives to full-scale adoption, where AI becomes an integral part of core operations and strategy. To scale AI successfully, organizations need a well-defined vision, strong leadership, and the necessary infrastructure to support widespread AI integration.

Scaling AI presents several challenges, starting with the need for a robust technical infrastructure. As AI applications grow, organizations must ensure they have the computational power and data storage capacity to handle the large volumes of data AI models require. This means investing in cloud computing services, AI platforms, and scalable data management solutions that allow AI algorithms to function efficiently as they expand across departments. Without this technical foundation, scaling AI will be hindered by performance bottlenecks and inefficiencies.

Another key challenge is organizational alignment. Scaling AI requires buy-in from all levels of the organization, from leadership to technical teams and business units. Fostering alignment ensures that everyone is on board with AI initiatives and understands their role in the broader strategy. Collaboration between departments is critical for integrating AI into different parts of the business and ensuring that AI solutions meet the diverse needs of the organization.

Governance and compliance become even more important as AI scales across the organization. Larger-scale AI deployments must adhere to ethical guidelines, maintain transparency, and comply with relevant regulations such as data privacy laws. This is particularly crucial in sectors like healthcare, finance, and retail, where AI has direct impacts on sensitive customer data and decision-making processes. Proper governance ensures that AI systems are developed and deployed responsibly, minimizing risks such as bias and misuse.

To effectively scale AI-driven innovation, organizations should first develop a scalable AI strategy. This involves creating a roadmap that outlines how AI-driven innovations will be scaled throughout the organization. The roadmap should include specific goals, timelines, and a clear allocation of resources to achieve widespread adoption. A scalable strategy helps ensure that AI initiatives are aligned with the organization's long-term objectives and that resources are deployed efficiently.

Investment in infrastructure and talent is also vital. Organizations must have the right tools, platforms, and expertise to scale AI successfully. This may involve upgrading data storage systems, adopting cloud-based AI platforms, or hiring specialists with AI and machine learning expertise. Building an internal AI talent pool ensures that the organization has the technical skills needed to support ongoing AI innovation and adoption.

In summary, scaling AI requires more than just expanding existing AI applications: it involves building a strong technical infrastructure, aligning organizational goals, investing in talent, and ensuring ethical governance. With the right strategy or, as depicted in Figure 19-1, a roadmap, businesses can successfully scale AI to drive innovation, enhance operational efficiency, and create sustainable growth across all areas of the organization.

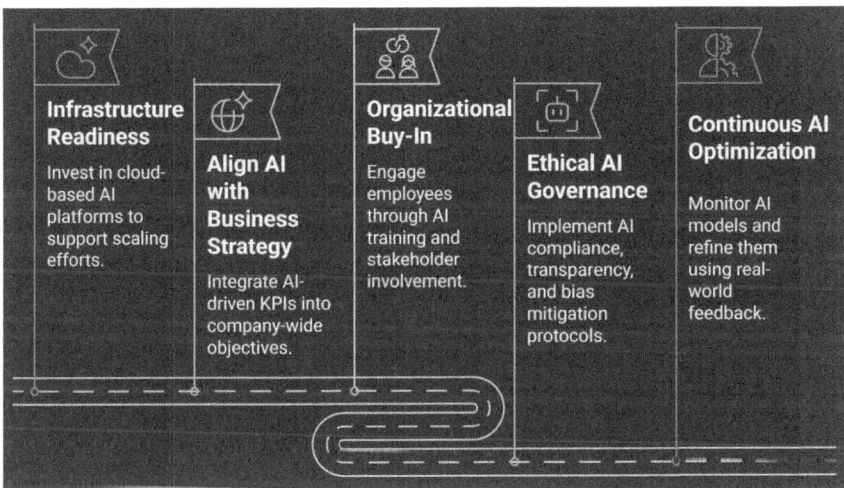

Figure 19-1: AI adoption roadmap

Key Takeaways for Leadership

AI-driven innovation is transforming industries and creating new business models that enable organizations to unlock growth opportunities and stay ahead of the competition. However, unlocking AI's full potential requires fostering an innovation mindset, investing in the right infrastructure, and scaling AI across the organization. Key recommendations for leadership include the following:

- **Identify high-impact areas where AI can drive innovation** and focus on developing AI-powered products, services, and business models that deliver significant value.

- **Encourage experimentation and cross-functional collaboration** to foster a culture of AI-driven innovation and ensure that AI solutions meet the needs of the entire organization.

- **Invest in infrastructure and talent** to support scalable AI adoption and ensure that the organization has the resources needed to leverage AI for long-term growth.

- **Implement governance frameworks** to ensure that AI systems are deployed ethically, transparently, and in compliance with regulations as they scale across the organization.

Ethical AI for Sustainability and Social Impact in Manufacturing

As sustainability becomes an urgent business imperative, manufacturing leaders must recognize that AI can be a powerful tool to drive both environmental and social impact—*if deployed ethically*. AI has the potential to reduce emissions, minimize waste, and support circular economies while also improving health, education, and equity in surrounding communities. But these benefits require leadership to ensure that AI initiatives are developed, governed, and scaled responsibly.

This chapter reframes sustainability as a leadership responsibility in AI ethics, combining the technical capabilities of AI with the strategic intent to deliver lasting environmental and social value. It builds on foundational AI ethics principles—fairness, transparency, accountability, and privacy—and applies them to sustainability use cases across energy optimization, resource management, social equity, and responsible supply chain design.

AI for Environmental Responsibility

AI technologies can dramatically improve environmental performance in manufacturing by optimizing resource use, reducing emissions, and enabling sustainable design. For example, Generative AI can simulate more energy-efficient product designs, and Agentic AI can autonomously adjust machine settings to minimize waste.

Leaders must ensure that these systems are trained on representative, high-quality data. In energy-intensive sectors like manufacturing and data centers, AI can optimize HVAC systems and load balancing to cut emissions—DeepMind's success in reducing data center cooling energy by 40% is one widely cited case. Figure 20-1 shows an example of the energy consumption reduction.

NOTE   To learn more about DeepMind, see "DeepMind AI Reduces Google Data Centre Cooling Bill by 40%" by Richard Evans and Jim Gao. It can be found at `deepmind.google/discover/blog/deepmind-ai-reduces-google-data-centre-cooling-bill-by-40`.

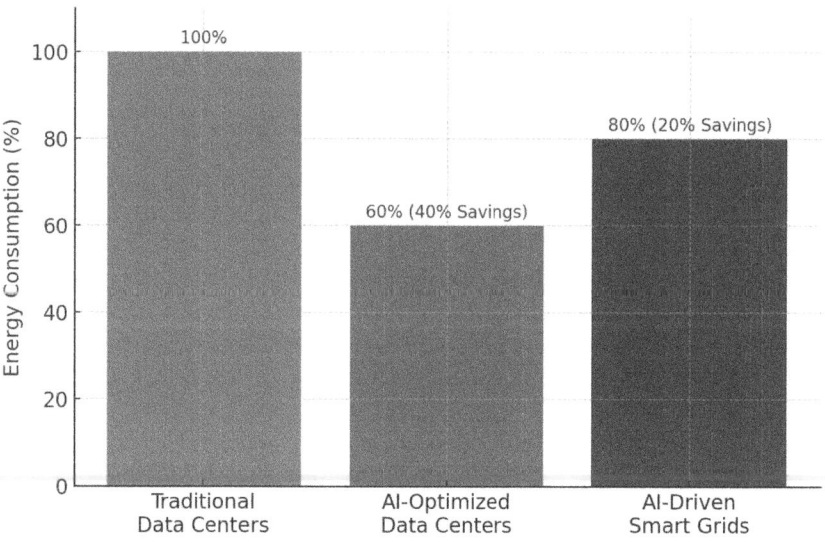

Figure 20-1: AI energy reduction in data centers

However, deploying such solutions in factories requires careful oversight to prevent excessive monitoring and opaque decision-making. Real-time optimization tools must include clear boundaries: when AI autonomously shifts energy usage or production schedules, human teams must understand *why*. Human-in-the-loop governance ensures that AI actions align with safety, compliance, and business priorities.

Enabling Circular Manufacturing Systems

In circular economy models, manufacturing aims to reduce waste and minimize resource utilization, so waste becomes a resource. AI can power this transformation by detecting material inefficiencies, recommending reuse pathways, or even reconfiguring logistics to support reverse supply chains. Generative AI models, for instance, can design components that use fewer virgin materials or are easier to recycle.

One of the key applications of AI in the circular economy is waste reduction and recycling. AI-driven systems can sort and process waste more efficiently, boosting recycling rates and minimizing the amount of waste sent to landfills. Machine learning algorithms can optimize waste management by identifying the most efficient recycling methods, analyzing waste streams, and recovering valuable resources that would otherwise be discarded. AI can also streamline waste management routes, making the overall system more efficient and environmentally friendly. AMP Robotics utilized AI and robotics to enhance recycling efficiency. Its systems can sort various materials such as paper, plastics, and metals with high accuracy, leading to a 10% increase in the amount of recycled material.

NOTE You can learn more about what AMP Robotics did in a report by the Ellen MacArthur Foundation called "Artificial Intelligence for Recycling: AMP Robotics." It can be found at www.ellenmacarthurfoundation.org/circular-examples/artificial-intelligence-for-recycling-amp-robotics.

Safeguarding Privacy in Environmental Monitoring

Many AI-driven environmental solutions rely on data from sensors, employee tracking systems, and industrial IoT networks. Although these tools are powerful, they raise serious privacy concerns. For example, using AI to optimize HVAC performance in a plant may involve tracking employee movement or workspace occupancy.

Leadership must set clear boundaries and enforce data minimization principles, collecting only what's necessary and ensuring that it is anonymized wherever possible. Consent mechanisms should be in place, particularly when personal or biometric data is involved. Ethical environmental AI solutions respect both the planet *and* the people working to protect it.

Transparent communication is key: manufacturing leaders should ensure that workers and stakeholders understand how sustainability data is collected, used, and safeguarded. Embedding privacy-by-design principles into all sustainability AI deployments enhances trust and long-term effectiveness.

AI for Social Impact

Beyond environmental gains, AI in manufacturing can enhance social equity, particularly in communities where plants operate. From workforce development to healthcare access, AI solutions can be deployed to uplift communities—*if guided by inclusive leadership.*

In this section, *social equity* refers to ensuring fair access to opportunities, resources, and outcomes—especially for individuals and communities that have been historically marginalized or disadvantaged. For example, AI-driven training programs can be used to reskill workers for the AI era, supporting long-term employability and reducing inequality. AI tools can also personalize safety training and language translation for multilingual teams, improving workplace accessibility and reducing risk.

Moreover, manufacturers can use AI to identify and address community health and infrastructure needs. A manufacturing company might partner with local governments to use AI to analyze traffic, pollution, and health service gaps near plants. However, ethical oversight is essential to prevent misuse or mission creep in community data.

AI should also be used to identify and correct bias in hiring, performance management, and promotion systems—areas where underrepresented groups have historically faced systemic inequities. By applying fairness algorithms and conducting regular audits, leadership can ensure that AI reinforces inclusion rather than replicating exclusion.

Aligning Sustainability with Governance Structures

To succeed, sustainability-focused AI must be embedded into governance frameworks, not treated as a one-off pilot. Ethical leadership involves setting clear sustainability objectives for AI, integrating them into enterprise risk management, and reporting progress transparently.

For example, a manufacturing firm might create an AI Sustainability Committee to oversee environmental and social impact programs. This group would review proposed AI systems through an environmental, social, and governance (ESG) lens: Does this tool reduce waste? Does it respect privacy? Does it support underserved populations?

International frameworks like ISO/IEC 42001 and regulatory movements like the EU AI Act increasingly require this kind of proactive oversight. Leading manufacturers are adopting internal standards that exceed compliance, embedding AI ethics into product lifecycle governance—from design and training to deployment and retirement. By doing so, organizations can ensure that AI is not only a driver of innovation but also a tool for achieving long-term sustainability goals. The following are tips for aligning AI with sustainability goals:

- **Set clear sustainability objectives for AI:** Establish specific, measurable goals for how AI will contribute to sustainability efforts. This might include reducing carbon emissions, minimizing waste, and improving access to essential services.

- **Integrate AI into corporate sustainability strategy:** Ensure that AI initiatives are part of the broader sustainability strategy. This alignment ensures that AI-driven solutions are prioritized and supported at the highest levels of the organization.

- **Measure and report on AI's impact:** Regularly measure and report on the impact of AI-driven sustainability initiatives. This transparency helps build trust with stakeholders and ensures that AI is contributing to the organization's overall sustainability goals.

Transparency also extends to stakeholders. Ethical leaders publish sustainability impact reports, including how AI contributed to key ESG metrics like emissions reductions and community engagement. In doing so, they position AI not just as a tool of efficiency but also as a lever for leadership.

Key Takeaways for Leadership

Sustainability and ethics are converging priorities—and AI is the bridge. Manufacturing leaders must treat sustainability-focused AI as a strategic capability, not a side project. The promise of AI-enabled energy optimization, circular production, and inclusive workforce development can only be realized when trust, fairness, and accountability are embedded at every level.

This requires more than good intentions. It demands governance, transparency, and investment. Ethical AI leadership means championing sustainability goals with the same rigor as productivity metrics. It means involving diverse voices in AI design. It also means recognizing that in the 1-degree world, environmental and social trust are core to competitive advantage.

When deployed ethically, AI enables manufacturers to innovate boldly, which results in cutting waste, elevating workers, and creating value that endures far beyond the factory floor. Key recommendations for leadership include the following:

- **Leverage AI to address environmental challenges**, such as reducing carbon emissions, optimizing resource usage, and minimizing waste.

- **Use AI to enable circular economy initiatives** by improving recycling processes, extending product lifecycles, and optimizing supply chains.

- **Implement AI-driven energy management systems** to improve energy efficiency and support the integration of renewable energy.

- **Integrate AI into CSR programs** to drive positive social impact, improve healthcare access, enhance education, and promote social equality.

- **Align AI initiatives with corporate sustainability goals**, and measure AI's contribution to environmental and social impact regularly.

Scaling AI: From Pilot Projects to Full-Scale Transformation

In today's rapidly evolving technological landscape, AI presents unprecedented opportunities to drive innovation, optimize operations, and unlock new business value. Many organizations begin their AI journey by experimenting with small-scale pilots and proofs of concept. These pilots typically focus on automating a single process, such as customer support through chatbots or implementing AI-powered recommendation engines for targeted marketing. These pilots allow organizations to test AI capabilities, gather data, and prove the return on investment (ROI) before making larger commitments.

However, although pilots can be insightful and help build the business case for AI, they only scratch the surface of AI's potential. The true value of AI lies in its ability to transform entire organizations by fully embedding AI into core business functions. The next challenge for leadership is scaling these AI innovations across the business, transforming AI from isolated use cases into a key component of everyday business operations and strategic planning.

Scaling AI is not merely about increasing the scope of deployments. It's about fully integrating AI systems into the core of the organization to drive long-term growth, innovation, and operational efficiency. To realize AI's full value, AI must evolve from experimental use cases into scalable systems that influence decision-making, improve productivity, and open new opportunities for innovation across all departments. For companies that successfully navigate the complexities of scaling AI, the payoff is a powerful lever for achieving competitive advantage and sustaining long-term growth.

From Pilots to Full-Scale AI Integration

Scaling AI from pilot projects to full integration offers transformative benefits beyond isolated improvements. AI at scale can automate routine tasks across departments, optimize workflows, predict demand, and even suggest strategic process improvements.

Many organizations find that AI's real impact is felt when AI technologies are fully integrated into business functions such as marketing, finance, operations, and supply chain management. This requires not just technical integration but also cultural and operational shifts. When AI systems are scaled across an organization, their efficiencies compound, allowing businesses to cut costs and improve productivity across multiple departments. The compounding benefits drive cost savings, faster decision-making, and increased innovation, all of which contribute to a competitive edge. In Figure 21-1, a four-step roadmap is shown that organizations can follow. By following this roadmap, organizations can ensure a structured and seamless transition from AI experimentation to full-scale adoption, maximizing AI's value and long-term impact.

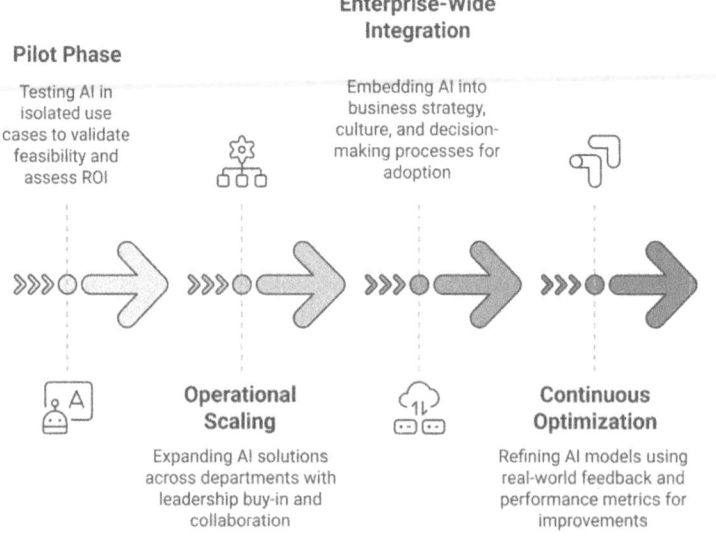

Figure 21-1: AI scaling roadmap

The Importance of Scaling AI

The full value of AI is realized when it moves beyond specific functions or departments to become a central part of decision-making, innovation, and long-term strategy. Here are key ways that scaling AI can transform an organization:

- **Operational efficiency:** One of the most immediate benefits of scaling AI is the significant improvement in operational efficiency. AI systems excel at automating repetitive, manual tasks—such as data entry, scheduling, and basic customer inquiries—freeing up employees to focus on more strategic work. When AI is scaled across multiple departments, these efficiencies multiply, helping organizations optimize workflows, predict demand, and reduce costs. AI can also continuously analyze operational data to identify bottlenecks and suggest improvements, creating a leaner and more agile operation.

- **Data-driven innovation:** AI thrives on data. Scaling AI across the organization enables businesses to harness the full power of their data. AI can analyze vast amounts of structured and unstructured data to uncover patterns, trends, and opportunities that may be invisible to human analysts. With scaled AI systems, organizations can tap into new markets, improve product development, and drive data-driven innovation. AI-powered analytics can transform customer feedback into actionable insights, refine marketing strategies, and generate entirely new products and services that resonate with emerging consumer needs.

- **Enhanced decision making:** AI's ability to process complex datasets makes it an invaluable tool for decision-making. Scaled AI systems provide real-time insights that help leaders optimize supply chains, forecast market trends, and make informed decisions faster. Whether AI is used to enhance financial forecasting, improve risk management, or deliver targeted customer experiences, scaled AI systems allow businesses to make smarter, data-driven decisions that drive better outcomes.

- **Competitive advantage:** Companies that successfully scale AI across their operations gain a significant competitive advantage. AI-powered organizations can respond more rapidly to changing customer needs, personalize their offerings, and optimize operations more effectively than their competitors. Additionally, scaling AI enables organizations to be more agile in responding to market changes, enabling them to stay ahead of trends and outperform rivals.

Leadership's Role in Scaling AI: Communicating Strategy and Ensuring Alignment

Scaling AI across an organization requires more than just deploying technology: it requires a comprehensive leadership strategy that ensures AI's integration into the company's core operations and long-term goals. For AI to succeed at scale, leaders must take ownership of the process, aligning AI initiatives with the organization's broader strategic vision and ensuring that the necessary infrastructure, talent, and resources are in place.

Communicating a Clear AI Strategy

One of the first tasks of leadership in scaling AI is to communicate a clear and compelling AI strategy that aligns with the organization's long-term goals. A successful AI strategy should answer key questions about the organization's future, including:

- **What is the role of AI in driving business outcomes?** Leaders must articulate how AI will impact key areas such as customer experience, operational efficiency, product development, and market expansion.

- **How will AI drive long-term growth and innovation?** Leaders need to outline how AI initiatives will support the organization's vision for the future, including which specific goals AI will enable, such as improved productivity, cost savings, and enhanced customer personalization.

- **What are the milestones for AI integration?** Clear communication about short-term wins and long-term goals helps employees understand the progress and purpose of AI initiatives. Setting timelines, objectives, and key performance indicators (KPIs) ensures that everyone stays aligned.

A well-communicated AI strategy helps establish trust and provides a roadmap for the workforce to follow. It also ensures that employees at all levels understand the broader context of AI adoption, including how it will benefit their roles and the organization as a whole.

Investment in AI Infrastructure

A critical aspect of leadership's role in scaling AI is ensuring that the organization has the necessary infrastructure to support AI at scale. AI systems thrive on data, and successful scaling requires a robust data ecosystem capable of managing large datasets, high-performance computing, and machine learning platforms. Leaders must invest in cloud computing and data storage solutions that provide the scalability and flexibility needed to handle vast amounts of data and support AI model execution. Cloud-based infrastructure ensures that AI projects can grow efficiently without limitations on computational power or data access.

In addition to storage and computing capabilities, organizations need AI tools and platforms that facilitate machine learning model development, data analytics, and integration across departments. Investing in AI development platforms ensures that teams have access to the necessary resources for continuous AI innovation. However, infrastructure alone is not enough— data governance must be a priority to maintain compliance with regulations, protect data privacy, and uphold ethical AI practices. Leaders should implement robust governance frameworks that ensure secure data management, transparency, and accountability in AI-driven decisions.

By making these infrastructure investments, leadership ensures that AI is scalable and sustainable. Establishing a strong AI foundation also signals a long-term commitment to AI adoption, which is crucial for gaining employee buy-in and fostering cross-functional collaboration.

Ensuring Alignment with Long-Term Goals

Scaling AI should be treated not as an isolated technology initiative but rather as an integral part of the organization's long-term strategic framework. Leaders must embed AI across various business functions, including operations, supply chain, customer service, and finance, to ensure organization-wide impact rather than isolated implementations. AI's role in business objectives must be clearly defined to create lasting value across departments.

Beyond business functions, AI initiatives should also align with the organization's core corporate values. If sustainability is a key focus, for example, AI can be leveraged to optimize energy consumption, minimize waste, and improve resource efficiency in production processes. Similarly, AI-driven personalization and automation should reflect the organization's commitment to customer-centric innovation.

To measure the success of AI adoption over time, leadership must establish long-term KPIs that align with the company's growth strategy. Success metrics could include increased revenue from AI-powered services, reduced operational costs through automation, and enhanced customer satisfaction from personalized AI-driven experiences. By embedding AI within the company's core objectives and values, organizations can ensure that AI contributes to long-term success rather than serving as a short-term technological experiment.

Change Management Strategies

Leadership must actively manage organizational change to overcome resistance and foster a collaborative AI culture. A key strategy is clear communication of AI's benefits. Leaders should consistently highlight how AI will enhance both organizational performance and individual employee roles. Demonstrating how AI reduces workloads, improves accuracy, and creates new opportunities can alleviate fears and encourage adoption.

Involving employees in AI initiatives is another crucial step. Employees are more likely to embrace AI if they are included in discussions about its integration into workflows. Organizations should seek employee input on how AI can enhance their work, giving them a sense of ownership and control over AI-driven changes.

Finally, reskilling and upskilling programs help employees feel prepared and empowered to work alongside AI. Offering training in technical skills and AI literacy ensures that employees can adapt to new technologies and remain valuable contributors to the organization. By investing in workforce development, organizations demonstrate a commitment to employee growth while reducing concerns over job displacement.

By addressing resistance, fostering transparency, and equipping employees with the necessary skills, organizations can successfully align their workforce with AI initiatives and create a culture that embraces technological innovation.

Key Strategies for Scaling AI

To navigate the challenges of scaling AI, organizations need a comprehensive, structured approach. The following strategies can help businesses scale AI effectively and unlock its full potential:

1. **Develop a clear AI vision and roadmap:** Scaling AI starts with a clear vision for how AI will drive long-term business goals. Leaders must articulate a detailed AI strategy that outlines the specific areas where AI will be integrated, the expected business outcomes, and the resources required for scaling. This vision serves as the framework for decision-making, helping leaders prioritize AI initiatives based on their potential impact.

2. **Invest in scalable data infrastructure and AI platforms:** A strong data infrastructure is essential for scaling AI. Organizations need cloud-based platforms, data storage solutions, and machine learning tools that are scalable, flexible, and secure. This infrastructure must handle large volumes of data, process complex algorithms, and deliver real-time insights.

3. **Build and retain AI talent:** To address the shortage of AI talent, organizations should invest in internal talent development through reskilling and upskilling. Additionally, leadership should foster a culture of learning to ensure the continuous development of AI-related skills. Cross-functional teams of data scientists, engineers, and business leaders should collaborate to ensure AI solutions are technically sound and strategically aligned.

4. **Foster a culture of innovation and collaboration:** Scaling AI requires a culture that embraces experimentation. Organizations should foster a culture where teams feel empowered to test new ideas and develop innovative solutions. Establishing AI labs and innovation hubs encourages cross-functional collaboration and supports continuous improvement.

5. **Establish strong governance and ethical frameworks:** Leadership should establish ethical guidelines to ensure that AI systems are governed responsibly. Governance frameworks should address issues such as data privacy, fairness, and transparency. Additionally, regular audits of AI systems will help ensure ethical compliance as AI scales.

CASE STUDY: SCALING AI IN A REGIONAL MANUFACTURING COMPANY

A local manufacturing company based in the Midwest in the United States initially implemented AI as part of a pilot program to optimize its production line efficiency. The pilot delivered promising results, with a 12% increase in production throughput and a significant reduction in equipment downtime due to AI-driven predictive maintenance. Encouraged by the success of the pilot, the company sought to scale its AI capabilities across the entire manufacturing process.

To achieve this, the company developed a comprehensive AI roadmap outlining how AI would be integrated into areas such as production planning, quality control, and supply chain management. The company invested in cloud-based AI platforms and upgraded its data infrastructure to ensure that its AI systems could process the large amounts of data generated by its factories and suppliers. This infrastructure allowed for real-time monitoring of equipment health, production metrics, and supply chain data, ensuring that AI insights could be acted on swiftly.

The company also launched a reskilling initiative, offering factory managers and machine operators training in AI-powered analytics and real-time data monitoring. This initiative enabled employees to work effectively alongside AI, using predictive insights to reduce downtime, optimize production schedules, and enhance quality control. By integrating AI-driven tools for defect detection and process optimization, employees were able to prevent costly quality issues and improve overall product consistency.

Scaling AI across the entire manufacturing process delivered a 20% reduction in operating costs by lowering equipment downtime and improving production efficiency. Additionally, AI's ability to predict and address potential bottlenecks in the supply chain led to a more streamlined operation, ensuring that production schedules remained on track and that raw materials arrived just in time for manufacturing needs.

By strategically scaling AI, the company not only reduced costs but also gained a significant competitive advantage in the manufacturing sector, increasing its production capacity and improving product quality without additional capital investments in equipment and facilities.

Key Takeaways for Leadership

The journey from AI pilot programs to enterprise-wide scaling is filled with challenges, but for organizations that navigate these complexities, the rewards are transformative. By embedding AI into the heart of their operations, businesses can unlock new levels of efficiency, innovation, and competitive advantage. Scaling AI is not the end of the journey but the beginning of a long-term transformation that will reshape industries, redefine roles, and create unprecedented opportunities for growth.

Key takeaways for unlocking long-term value through AI at scale include the following:

- **Scaling AI unlocks enterprise-wide value:** Moving beyond pilots to integrated AI systems drives efficiency, innovation, and sustainable competitive advantage across operations.

- **AI is a long-term transformation, not a one-time initiative:** Embedding AI into core business functions sets the stage for ongoing evolution in how work is performed, decisions are made, and value is created.

- **Leadership is critical to success:** Visionary leaders must champion AI initiatives, invest in infrastructure and talent, and foster a culture of innovation and adaptability.

- **Ethics and societal impact must be part of the strategy:** As AI reshapes industries and job roles, leaders must address its broader implications—ensuring responsible deployment that upholds fairness, accountability, and social good.

- **The journey continues:** Scaling AI is only the beginning. Organizations must remain agile, continuously aligning AI with business goals and emerging opportunities while staying vigilant about its ethical dimensions.

As organizations scale AI and reap its benefits, it is equally vital to consider the broader ethical and societal implications. The adoption of AI will not only transform businesses but also reshape the fabric of work and society. In the next chapter, we'll explore the ethical dilemmas and responsibilities that accompany this transformation.

Leading in the AI-Driven 1-Degree World: Ethics, Workforce Transformation, and Global Strategy

As AI becomes deeply embedded in manufacturing, leaders face a dual mandate: harness AI for innovation and efficiency while upholding ethics and preparing their workforce for unprecedented change. In this 1-degree world—a hyper-connected global ecosystem where even a 1-degree shift in strategy and values can ripple worldwide—manufacturing executives must lead with foresight and integrity. This capstone chapter explores how ethical AI governance, talent transformation, and strategic integration of Generative and Agentic AI come together to reshape global manufacturing. It builds on earlier themes of visionary leadership and responsible innovation, serving as a call to action for leaders to navigate the future of manufacturing in an AI-driven era.

Ethical AI Governance in Manufacturing

As AI scales across manufacturing, leaders must ensure that its use is fair, transparent, and accountable. Common ethical challenges—such as algorithmic bias, opaque decision-making, and unclear accountability—can erode trust if left unaddressed. For example, an AI system that optimizes supply chains but unintentionally favors certain regions could reinforce inequities. Proactive governance anticipates these risks before they harm performance and reputation. We covered this in more depth in Chapter 10.

Leading organizations now conduct regular AI audits to identify and mitigate bias, using cross-functional teams to review training data, model logic, and outcomes. These practices are bolstered by formal data governance frameworks and ethics committees that oversee responsible AI deployment. In manufacturing, such committees ensure that quality control systems and HR algorithms comply with privacy laws and fairness standards. Ethics boards are also being established to vet AI solutions before launch, aligning them with organizational values and regulations. Transparency—through published AI use policies and ethics reports—helps build trust with employees, customers, and regulators alike.

Manufacturers are increasingly aligning governance with global standards such as the EU's AI Act and ISO/IEC 42001. Companies like i-PRO have launched comprehensive frameworks, integrating ethics into each stage of AI product development, from design to deployment. These systems ensure that responsible AI is embedded in daily operations—not treated as an afterthought. Effective governance includes ethical guidelines for developers, bias mitigation protocols, human-in-the-loop controls for critical decisions, and clear escalation paths. Leadership must assign accountability so that AI-related decisions can be monitored and corrected when necessary. In the hyper-connected 1-degree world, trust and transparency in AI aren't just good practice—they're competitive advantages.

NOTE For more information on what i-PRO has done, see "i-PRO Establishes Pioneering AI Governance Framework and Ethics Committee," which can be found at `i-pro.com/products_and_solutions/en/` `surveillance/newsroom/i-pro-establishes-pioneering-ai-` `governance-framework-and-ethics-committee`.

Transforming the Workforce for an AI-Enabled Future

Whereas governance provides guardrails for AI, the workforce is the engine driving AI's success. AI's rise is reshaping job roles across manufacturing, creating both excitement about productivity gains and anxiety about job security. Leadership's mandate is to navigate this workforce transformation in a way that empowers employees rather than alienates them. This begins with a commitment to upskilling and reskilling at all levels of the organization. As AI takes over repetitive tasks—from data entry in supply planning to visual inspection on production lines—employees must be enabled to move into higher-value roles that leverage uniquely human skills. For example, if AI-powered vision systems handle initial quality checks, technicians can be trained to focus on complex problem-solving, equipment tuning, and AI system oversight. Organizations should identify roles most susceptible to automation and proactively provide training pathways for affected staff to transition into new positions. This may mean a machine operator learns to manage a fleet of collaborative robots (cobots) or a warehouse supervisor learns to interpret analytics from an AI-driven logistics platform.

Actionable upskilling strategies include partnerships with educational institutions and in-house training programs. Some manufacturers are partnering with technical schools to offer certification courses in AI, robotics, or data analytics for their employees. Others have launched internal AI academies where workers can learn skills like programming simple AI scripts or maintaining automated equipment. The goal is not to turn every worker into a data scientist but to ensure that each employee has the digital literacy and AI awareness needed for their evolving role. For instance, a maintenance mechanic could learn to work with AI predictive maintenance systems, interpreting AI alerts about machine health and deciding on repairs. A procurement specialist might upskill in using AI forecasting tools for demand planning.

By investing in people, leaders signal that AI is a tool for *empowerment*, not a replacement. This investment pays off: companies that reskill staff often see higher productivity and morale. One global consumer goods company that extensively reskilled its workforce in AI tools saw improved efficiency, as reskilled workers could manage AI-driven supply chains and predictive maintenance systems with confidence.

Beyond skills, leaders must address the mindset and cultural aspects of workforce transformation. Introducing AI can spark fear and uncertainty: employees may worry about whether AI will make their experience irrelevant. Establishing psychological safety in this context is paramount. *Psychological safety* means creating an environment where employees feel free to voice concerns, ask questions, and even fail in the course of learning without fear of ridicule or retaliation. Leaders can cultivate this by openly discussing AI initiatives and their implications. They should encourage teams to share feedback on pilot projects and report issues or ethical concerns they observe in AI behavior. For example, if a factory worker notices that an AI scheduling system consistently under-schedules a certain shift, they should feel safe to flag this potential bias or error.

According to research at Google (see "Psychological Safety and AI Adoption," Bloomreach, Oct. 2024, www.bloomreach.com/en/blog/psychological-safety-and-ai), teams with higher psychological safety are more effective, especially when tackling new and complex challenges. In manufacturing, this translates to more successful AI adoption: when employees trust that their input is valued, they become active collaborators with AI systems rather than passive observers. They will more readily experiment with AI tools, accelerating innovation.

One way to foster this safety is through transparent communication about AI's purpose and impact. Leaders should clearly articulate why the company is adopting AI: for example, "We are deploying AI in our quality control to reduce defects and free up your time for more advanced troubleshooting." They should also be honest about the changes to job functions and the support being provided. Regular town halls, Q&A sessions, and demonstration workshops can demystify AI. As employees see AI in action and understand its limitations, fear of the unknown diminishes. Additionally, success stories of human–AI collaboration from within the organization (or industry) can be shared to highlight positive outcomes. Perhaps an engineer teamed with a generative design AI to create a new product component, or a supply planner used an AI agent's recommendation to prevent a stockout—celebrating these wins helps others see AI as an ally.

Crucially, manufacturing leaders should promote human–AI collaboration as the new workforce model. Rather than framing it as humans *versus* machines, the narrative should be humans *augmented by* machines. In many roles, AI serves as a decision support tool or a productivity booster, with final judgment still resting on human expertise. For example, AI may compile and analyze production data across global plants, but human managers use those insights to make strategic decisions about process improvements. This collaborative model extends to the factory floor: cobots (collaborative robots) working alongside people in assembly tasks amplify human capacity—the robot handles heavy lifting or precision tasks, and the human handles the finicky assembly and quality tweaks. Emphasizing this partnership model can alleviate fears and actually boost engagement, as employees see that their role is essential in guiding and validating AI. A welder who supervises an AI-driven robotic welding arm, for instance, becomes a mentor and controller for the AI, intervening when complex welds require a creative touch.

To solidify a collaborative, AI-ready culture, leaders can implement initiatives that reward learning and innovation. Continuous learning should be incentivized—perhaps through badges, certifications, and career advancement for those who master new AI-related skills. Some companies hold AI innovation challenges and hackathons for their employees, sparking creative ideas for AI use and signaling that experimentation is welcome. Establishing an *AI-first* culture—one that encourages data-driven decision-making and openness to new technology—often starts with leadership behavior. When leaders themselves use AI tools (e.g., using AI analytics in their strategy meetings or leveraging a Generative AI model for brainstorming ideas), they set a powerful example that cascades down. Over time, this culture of innovation and trust helps the workforce not only adapt to AI but actively *drive* AI initiatives. In the 1-degree world of manufacturing, where adaptability is key, an engaged and skilled workforce becomes a company's greatest asset.

Integrating Generative and Agentic AI into Manufacturing Functions

With governance frameworks and a skilled workforce in place, manufacturing leaders can begin integrating AI into core functions. Two complementary forms—Generative AI (which creates) and Agentic AI (which acts)—are particularly transformative when applied to areas like supply chain, quality control, and product design. The following illustrate how manufacturing leaders can apply these dual AI capabilities to critical operational domains—each offering unique opportunities to streamline processes, enhance agility, and unlock new value at scale:

Supply Chain and Logistics   Generative AI improves forecasting by simulating scenarios based on historical data, and Agentic AI enhances agility by making autonomous decisions in real-time. For instance, generative models may analyze seasonal patterns to create demand plans while AI agents reroute shipments in response to port delays or changing conditions. Some manufacturers already use autonomous procurement agents that negotiate with suppliers and dynamically allocate orders based on cost or risk.

Integrating these capabilities requires connecting supplier, inventory, and sales data into an AI-driven control tower. A generative model can recommend optimal inventory distribution while an agentic system autonomously balances stock levels and places replenishment orders. During crises like COVID-19, companies with AI-augmented supply chains adjusted faster, minimizing disruption. Leaders should start small—pilot fleet routing or demand forecasting AI—and then scale successful initiatives, always ensuring that human oversight defines the parameters and intervenes when needed.

Quality Control and Maintenance   AI has long supported machine vision in defect detection, but Generative AI takes this further by producing synthetic defect data to improve training. It also enables digital twin simulations to predict failures and optimize testing processes. Predictive maintenance becomes proactive: generative models forecast degradation scenarios, prompting action weeks in advance.

Agentic AI autonomously monitors production lines, adjusts settings in real time, and halts operations when critical defects spike. It doesn't just flag problems—it acts on them. Together, generative and agentic capabilities enable systems that both anticipate and respond. For example, an AI may identify a novel defect, classify it, suggest root causes, and then trigger containment actions automatically.

Smart factories like those operated by Schneider Electric use AI to optimize energy and reduce defects, improving both sustainability and profitability. Generative design algorithms have cut component weight and development time by 30–50%, increasing both quality and efficiency. Leaders should target quality and maintenance functions for early AI adoption, where returns are tangible and customer impact is direct.

Product Design and Engineering   Generative AI revolutionizes R&D by proposing optimized part designs based on performance criteria. Instead of manually testing each idea, engineers receive dozens of high-performing options instantly. In one case, a manufacturer used Generative AI to redesign a bracket, reducing weight by 26% and cost by 8%. These capabilities improve competitiveness across automotive, aerospace, and consumer product sectors.

Beyond structural optimization, Generative AI can generate schematics, user manuals, and even localized design variations. For example, an AI may turn a 3D model into visual assembly instructions tailored for factory workers—speeding the transition from concept to production.

Agentic AI supports the development process through intelligent coordination. Project management agents can track design tasks, flag risks, and auto-schedule meetings after failed tests. In the longer term, they may autonomously handle routine engineering changes—like sourcing substitute materials within defined constraints—and improve responsiveness and efficiency.

To begin, leaders should support R&D teams in piloting AI tools on manageable projects, integrating generative models into one component design or deploying an agentic assistant for a specific product line. Human oversight remains essential: AI may optimize for simulation metrics but fail to meet usability, cost, and aesthetic standards. The best results come when human creativity guides AI's computational horsepower.

Together, Generative and Agentic AI compress innovation timelines, enhance execution, and enable real-time decision-making across global operations. In the 1-degree world, an idea generated in one region can be evaluated, refined, and implemented in another within days—blurring the line between design and production and creating a new era of industrial agility.

The 1-Degree World: AI and the Global Manufacturing Ecosystem

Manufacturing today operates as a tightly interwoven global network: a change and innovation in one place can be felt everywhere almost immediately. This reality is what we call the *1-degree world*, evoking the idea that we are all just 1-degree apart in a massive ecosystem. AI is a driving force in tightening these global interconnections. For manufacturing leaders, embracing the 1-degree world means recognizing that AI-enabled globalization brings both opportunity and responsibility on a worldwide scale.

A significant aspect of the 1-degree world is the rise of truly global supply chains that behave like cohesive organisms. AI systems allow a level of real-time coordination and visibility across continents that was unimaginable in the past. A centralized AI platform can monitor factory outputs in Asia, inventory levels in Europe, and customer demand signals in the Americas simultaneously, adjusting plans so that the entire network functions in sync. In effect, AI shrinks the degrees of separation: a factory downtime event in one country can automatically trigger mitigations in another, thanks to AI. This improves resilience: for example, if political unrest delays a shipment from one region, an AI system may swiftly redistribute production to other plants or switch to alternate suppliers, communicating the changes instantly to all stakeholders. Leaders must design their organizations to leverage this connectivity. That can mean standardizing data and processes across global facilities so they feed into a unified AI "brain" or investing in IoT sensors and cloud infrastructure that connect every machine and supply chain node into a shared data ecosystem.

The 1-degree world also implies that ethical and sustainability standards are now global concerns. With AI magnifying the reach of manufacturing decisions, a lapse in one locale can tarnish a company's reputation everywhere. For instance, if an AI system in a plant inadvertently violates labor guidelines or causes environmental harm, news can spread globally in a flash. Therefore, the ethical AI governance discussed earlier must extend across borders: companies need consistent AI policies and ethics training in every region they operate. Some firms are adopting global AI ethics charters, ensuring that their use of AI in any country meets the same high standards for fairness and transparency. Cross-cultural and cross-functional ethics boards can include representatives from different geographies to address local nuances while enforcing global principles.

In a similar vein, AI is a powerful enabler of global sustainability efforts in manufacturing. Sustainability is inherently a cross-border issue: reducing carbon emissions in a factory contributes to mitigating climate change for everyone. AI helps by optimizing energy usage, reducing waste, and improving resource efficiency on a global scale. We see examples of this in smart factories where AI-driven systems cut energy consumption by dynamically adjusting lighting, heating, and machine performance, all coordinated through a central platform. When multiplied across dozens of facilities, these efficiencies make a substantial contribution to global climate goals (literally helping keep the world closer to a 1-degree temperature rise scenario, aligning with climate targets). Manufacturing leaders should tie AI initiatives to their sustainability strategies: for instance, using AI to route shipments in ways that minimize carbon footprint, or leveraging Generative AI to design lighter products that require fewer materials and less fuel to transport. By doing so, they not only save costs but also respond to the growing demand from consumers and governments for environmentally responsible production.

Another characteristic of the 1-degree world is the democratization of technology and knowledge. AI tools and insights can be instantly shared and replicated. A breakthrough machine learning model developed in one factory (say, an algorithm that dramatically improves yield) can be uploaded to the cloud and deployed to every other factory that same week. This means competition can be fiercer—advantages may only offer a brief edge before others catch up—but it also means rapid progress if embraced. To stay ahead, organizations should promote global collaboration through AI. This may involve global centers of excellence where data scientists in different countries jointly develop AI solutions, or encouraging teams to contribute to a common AI knowledge base. It requires interoperable systems and a willingness to break down silos between regions. The leaders who succeed will be those who can orchestrate their entire global enterprise almost like a single, borderless operation fueled by data and AI.

In summary, the 1-degree world intensified by AI presents manufacturing leaders with a dual reality: unparalleled connectivity and reach, paired with amplified accountability. Every strategic decision about AI—what data to use, which processes to automate, how to involve the workforce—has worldwide implications. But this also means even a small positive change (a "1-degree" improvement) can scale up through the network to yield a massive impact. The leadership challenge is to ensure that those impacts are positive: ethically, socially, and economically.

Conclusion: A Leadership Call to Action

Standing at the nexus of technology, people, and global strategy, manufacturing leaders are charged with steering their organizations through the next industrial revolution—one powered by AI and defined by a 1-degree world of interconnection. This final chapter has underscored that success in this era requires more than just adopting new technologies; it demands holistic leadership that blends ethics, talent development, and innovative strategy.

Leaders must champion ethical AI not as a box-checking exercise but as a core business practice that builds trust and longevity. Establishing governance frameworks, audits, and ethics boards is part of this, but so is leading by example—demonstrating an unwavering commitment to fairness and accountability in every AI project. When employees and partners see leaders prioritizing responsible AI (even if it means slowing a deployment to get it right), it sets the tone for an organization that values *doing the right thing* as much as *doing the fast thing*. In turn, this ethical backbone supports innovation: teams are more likely to experiment with AI when clear guardrails are in place, knowing they won't stumble into ethical quagmires.

At the same time, leaders must be the architects of workforce transformation. This involves communicating a vision where AI elevates human roles and then making that vision real through action: funding training programs, rewarding learning, and redesigning jobs to incorporate AI. A leader's willingness to invest in their people, even though the uncertainties of AI adoption, sends a powerful message of trust and a shared future. It builds the psychological safety and buy-in needed for the organization to truly embrace AI. And in practical terms, a workforce that continuously grows in skill and mindset is a workforce that will maximize AI's benefits. They will find creative uses for Generative AI, fine-tune Agentic AI systems on the ground, and ensure that technology implementations actually work in practice.

Finally, embracing the AI-driven future is about mindset. The leadership mandate in the AI era is to instill a sense of optimism and purpose about technology. Just as previous generations of industrial leaders led their people through quality revolutions and lean transformations, today's leaders must rally their organizations around digital transformation with AI. This chapter, as a capstone, reinforces that those who take a proactive and ethical approach to AI—who see it not just as a cost-cutter but as a generator of new value—will be the ones to thrive and define the next chapter of manufacturing. The AI-driven future of manufacturing is not about replacing people; it's about enabling people to work smarter, solve complex problems, and drive innovation at a scale never before possible. By investing in Generative AI for creativity, Agentic AI for efficiency, and, above all, in the culture and talent to use these tools wisely, manufacturing leaders can create a resilient, innovative, and inclusive industrial ecosystem.

The call to action is clear: lead boldly and responsibly. The tools and technologies detailed in this playbook—from ethical AI governance models to workforce enablement strategies and advanced AI integrations—are at the disposal of every manufacturing leader. It falls upon you to apply them in your context, to experiment and learn, and to share your successes and lessons in that 1-degree-connected global community of manufacturers. By doing so, you won't just react to the future: you will actively shape it. In the age of AI and the 1-degree world, the companies that pair innovation with integrity and automation with human-centric growth will set the standard for the next generation of industrial excellence. Let that be the legacy of your leadership.

Index